Diversity and Inclusion Research

Series Editor
Thomas Köllen, Institute of Organization and HRM
University of Bern
Bern, Switzerland

The book series 'Diversity and Inclusion Research' examines the facets of diversity in a variety of contexts, as well as approaches to and perspectives on diversity and inclusion. It presents organizational research on workforce diversity, and covers diversity and inclusion-related issues within communities, cities, regions, nations, and societies.

The series welcomes functional and intersectional perspectives on diversity, as well as constructivist perspectives on processes of "doing" or "performing" diversity. In terms of inclusion, it welcomes research on broader initiatives (such as generalized diversity management approaches), as well as on practices of inclusion that focus on specific dimensions of diversity; these can include age, appearance, ethnicity, disability, gender, gender identity, nationality, race, religion, sexual orientation, sex, or any other characteristic or demographic in which persons can differ from each other, or, conversely, which they share.

The series publishes research from the broad range of academic disciplines within the social sciences (e.g. economics and business administration, linguistics, political science, history, law, psychology, sociology, geography, and education) and humanities (e.g. philosophy and religion). It strongly encourages interdisciplinary and transdisciplinary research, as well as critical perspectives on diversity and inclusion-related issues. By welcoming research from a variety of sectoral, national, and cultural contexts, the series will make an essential contribution to shaping the academic discourse on diversity and inclusion.

Marc Helmold • Malte Martensen

Leadership in Neurodiversity and Neurodivergence

Cases, Best Practices and Recommendations for Leaders and Organizations

Springer

Marc Helmold
Management
IU International University of Applied Sciences, Campus Berlin
Berlin, Germany

Malte Martensen
Human Resources
IU International University of Applied Sciences, Campus Bremen
Bremen, Germany

ISSN 2662-5997 ISSN 2662-6004 (electronic)
Diversity and Inclusion Research
ISBN 978-3-032-09342-4 ISBN 978-3-032-09343-1 (eBook)
https://doi.org/10.1007/978-3-032-09343-1

This Springer imprint is published by the registered company Springer Nature Switzerland AG
The registered company address is: Gewerbestrasse 11, 6330 Cham, Switzerland

Acknowledgements

Neurodiversity and Neurodivergence describe the idea that people experience and interact with the world around them in many different ways; there is no one "right" way of thinking, learning and behaving, and differences are not viewed as deficits. There are estimates and surveys that assume 15 to 20 percent of all employees are neurodiverse. Depending on the definition, this includes people with ASD, ADHD, dyslexia and dyscalculia, autistic people, but also hypersensitive or highly gifted people. In contrast to neurotypical people, they have a different way of absorbing and processing information or expressing feelings than the majority.

The word neurodiversity refers to the diversity of all people, but it is often used in the context of autism spectrum disorder (ASD), ADHD, dyslexia or other learning disabilities.

The authors show in a pragmatic way how neurodiversity and inclusive leadership will shape the new world of work and why inclusion is a significant requirement for any organization. Moreover, it is shown how enterprises can successfully transform into a neurodiverse organization.

The book closes the gap between theory and practical science by giving recommendations for organizations and leaders on how to deal with neurodiversity and neurodivergent employees.

The book covers how to attract, recruit and engage neurodiverse talent and provides guidance for Leaders on how to adapt HR practices and processes to ensure that all employees can reach their full potential.

July, 2025 – Bremen, Berlin

Marc Helmold

Malte Martensen

List of Acronyms and Abbreviations

ADHD	Attention deficit hyperactivity disorder
AGG	Allgemeine Gleichbehandlungsgesetz
AI	Artificial intelligence
AL	Active listening
AS	Autism spectrum
ASD	Autism spectrum disorder
DCD	Developmental co-ordination disorder

EDI	Equality, diversity and inclusion
EI	Emotional intelligence
EY	Ernst & Young
NCoE	Neuro-Diverse Center of Excellence
NEDBELS	Neurodiversity between law and science
SPD	Sensory processing disorder
TS	Tourette syndrome
UDHR	Universal Declaration of Human Rights
VR	Virtual reality

Contents

About the Authors

Marc Helmold Prof. Dr. Marc Helmold (M.B.A.) is full-time Professor at IU Internationale Hochschule in Berlin. He teaches Bachelor, Master and M.B.A. courses in Business Management, Operations Management, Lean Management, Quality Management and Leadership.He had several positions in executive leadership positions at OEMs in the automotive, railway and aviation industry. Since 2016 he has been Professor at the IU Internationale Hochschule in Berlin (Germany) and conducts Lean, Quality and Leadership workshops in his own consultancy.

Malte Martensen Professor Malte Martensen, Dr. rer. pol., teaches and conducts research at IU International University of Applied Sciences in Bremen and Berlin. He specializes in leadership, entrepreneurship, organizational psychology, human resources management and the digitalization of work.In addition to his academic responsibilities, Malte consults organizations of various sizes and industries on issues of strategic HRM. He is also a speaker, trainer and certified coach, as well as a recognized author of numerous academic publications.

List of Figures

List of Tables

Neurodiversity and Neurodivergence: Subject and Definition

1

1.1 Neurodiversity in Society and Organizations

1.1.1 What Is Neurodiversity

Neurodiversity refers to the natural variation in how people and individuals think, process information, and experience the world. Neurodivergent individuals, such as those with autism, ADHD, autism, dyslexia, or Tourette's syndrome, bring unique strengths to the workplace, but their inclusion remains still limited (Amstrong, 2011). Neurodiversity refers to the virtually infinite neurocognitive variability within Earth's human population. It points to the fact that every human has a unique nervous system with a unique combination of abilities and needs. The importance of understanding neurodiversity is evidenced by the fact that up to 20% of the world's population exhibits some form of neurodivergence, with dyslexia being the most common type of a neurodivergent condition (Amstrong, 2011). Neurodiversity is made up of the terms "neuro" (nerves) and "diversity." This refers to the assumption and attitude that neurobiological differences in the brain are part of the spectrum of our development and do not represent a disorder or illness. For example, attention deficit hyperactivity disorder (ADHD) and autism are not seen as developmental disorders, but as neurological diversity. This is intended to shift the focus from concepts such as illness and deficit to diversity and individuality. According to the concept of neurodiversity, people with ADHD, autism, and other "learning disorders" do not have a disability or disorder. Instead, they simply reflect a different type of functioning of the human brain—an expression on a broad (normal) spectrum.

Neurodiversity is a relatively new phenomenon with the term neurodiversity being coined by Judy Singer, an Australian sociologist, in 1998, to describe people whose cognitive differences affect how their brains work (Amstrong, 2011). Neurodiversity addresses and normalizes human neurocognitive heterogeneity to

M. Helmold, M. Martensen, *Leadership in Neurodiversity and Neurodivergence*, Diversity and Inclusion Research, https://doi.org/10.1007/978-3-032-09343-1_1

promote acceptance and inclusion of neurominorities (e.g., learning disabilities, attention disorders, psychiatric disorders, and more) in contemporary society.

Since the 1990s, the neurodiversity movement has been directed at raising awareness and appreciation for the diversity in human cognition and breaking down structures of exclusion. However, within the neurodiversity movement, there is an ongoing debate about what self-advocacy means and the aspired social and political outcomes. Some argue that neurological differences should be normalized and accepted. Others argue against normalization and seek acknowledgment of the disability or illness related to their neuronal atypicality. These two opposite approaches correspond (respectively) with the social and (classic) medical models of disability. According to the medical model, disability is caused by a dysfunction of the individual and requires treatment or a cure. The social model distinguishes between disabilities and impairments, focusing on social barriers that limit the accessibility of certain individuals, ultimately leading to disability. Given that humans are complex, with multiple traits underlined by neurological structures and mechanisms, neurodiversity is multidimensional, and the distinction between "neurotypical" and "neurodivergent" is not straightforward. For example, left-handedness is related to atypical brain lateralization that characterizes only 10–15% of the global population and therefore is a form of neurodiversity. While left-handed people suffered discrimination in the past due to their atypicality, this has changed over the years. Today, left-handedness is no longer perceived as sinister and usually does not lead to social exclusion. Hand dominance has been normalized so much that left-handed people are not part of the neurodiversity movement. The case of left-handedness is an example of the link between neurodiversity and dynamic social norms. In many aspects, this normalization process is what the neurodiversity movement is seeking today (Marks, 1999).

Figure 1.1 depicts the different types linked to neurodiversity. Most commonly recognized forms of neurodiversity are ADHD, autism, and the Tourette syndrome. Additionally, there are mental health elements, acquired neurodiversity, dyslexia, dyscalculia, and dyspraxia.

Organizations with neurodivergent employees can benefit through hyperfocus, innovative thinking, resilience, sensory awareness, verbal skills, creativity, and authenticity.

1.1.2 Neurodiversity, Neurodivergence, and Neurotypicality

The neurodiversity paradigm is a framework for understanding the human brain function that recognizes the diversity within sensory processing, motor abilities, social comfort, cognition, and focus as neurobiological differences. This diversity falls on a spectrum of neurocognitive differences. The neurodiversity paradigm argues that diversity in neurocognition is part of humanity and that some neurodivergences generally classified as disorders, such as autism, are differences and disabilities that are not necessarily pathological (Kapp, 2020). The term neurodiversity is most commonly now used to refer to a group that encompasses the full spectrum

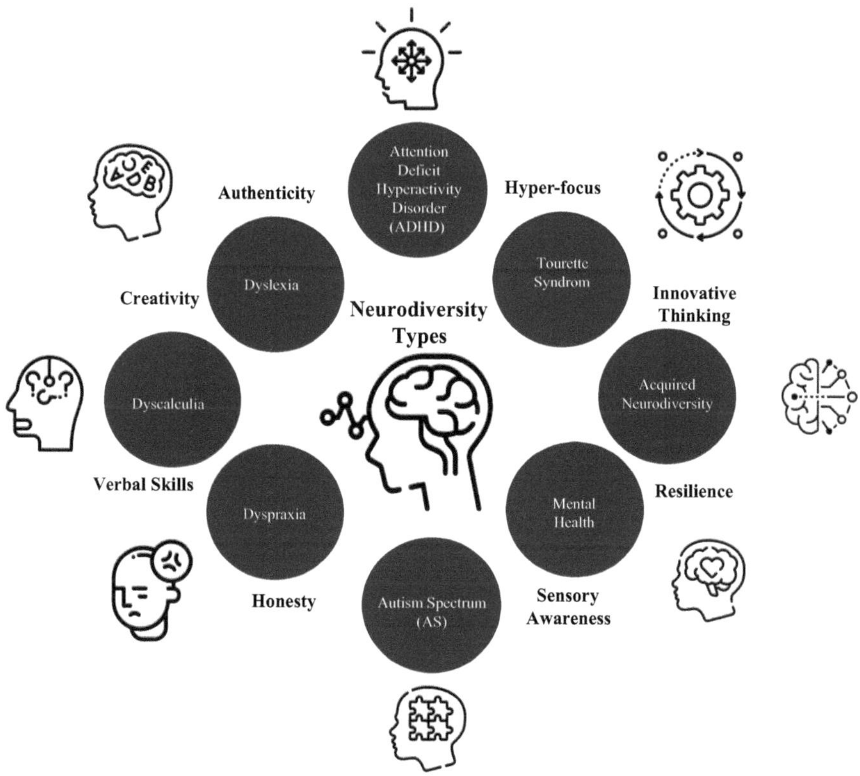

Fig. 1.1 Types of neurodiversity. Source: Author

of brain differences and is made up of both neurodivergent and neurotypical individuals.

The word neurodiversity first appeared in publication in 1998 in an article by American journalist Harvey Blume, as a portmanteau of the words neurological diversity to describe the growing concept of a natural diversity in humanity's neurological expression (Blume, 1998). Neurotypicality is used to describe people whose brain functions, ways of processing information, and behaviors are seen to be standard. Neurotypical individuals are those who fall within the average range of functioning and thinking.

1.2 Types of Neurodiversity

1.2.1 Attention Deficit Hyperactivity Disorder (ADHD)

The attention deficit hyperactivity disorder (ADHD) is one of the most common mental disorders affecting children. Symptoms of ADHD include inattention (not being able to keep focus), hyperactivity (excess movement that is not fitting to the setting), and impulsivity (hasty acts that occur in the moment without thought). ADHD is a neurodevelopmental condition affecting brain structure and neurotransmission: the way in which messages are communicated around the brain and different areas of the brain are activated. ADHD is thought to be caused by a complex mix of environmental and genetic factors, but is a strongly hereditary condition. ADHD is present from childhood, but an increasing number of adults are now being diagnosed with ADHD for the first time, having been "missed" when they were younger and the condition was not as well understood as it is now. ADHD has three core symptoms that affect people with it to different degrees:

Inattention
- Difficulties with concentration, short-term and working memory
- Difficulties with planning and getting started (activation)
- Difficulty with organization and losing things
- Easily distracted by small things that others would not notice

Impulsiveness
- Acting or speaking on the spur of the moment without thinking through the consequences
- Difficulty controlling emotions

Hyperactivity
- While adults with ADHD are usually much less active than children with ADHD, they may still have symptoms such as restlessness and the need to tap or fidget. Some people are diagnosed with attention deficit disorder, without hyperactivity; this is particularly the case for girls and women

For leaders, it is important to understand the challenges faced by employees with ADHD. Leaders and work teams must make an effort to understand their perspective and offer support when needed. Employees with ADHD are more likely to suffer from chronic stress and burnout. These problems can be counteracted concretely. On the other hand, ADHDers prove to be particularly reliable in emergency situations. In case of special interest (positive hyperfocusing) and the need for innovation, many ADHDers grow beyond themselves and demonstrate above-average performance potential (Lachemaier, 2023). As positive as these areas are, they also harbor risks for misunderstandings and undesirable developments. Recognizing and concretely preventing these risks are of great benefit to both the affected persons and their employers.

ADHD can work better when there is not too much happening around them all the time. That's why they need a silent workplace instead of loud open-plan offices, or at least temporary options. If there are too many stimuli bombarding employees with ADHD disorder all the time, it can have a significant negative impact on the team performance. Leaders must ensure that these employees with ADHD have everything they need for a quiet and concentrated working atmosphere. In addition, they have to initiate sufficient breaks. Shorter time intervals also include strategically sensible breaks (Lachenmaier, 2023). The constant exposure to all kinds of external and internal stimuli can be extremely exhausting. That's why managers should definitely give their employees with ADHD enough time-outs. This can be done, for example, by setting up a specially designed (small) break room. Organizations can then design it in such a way that no unnecessary stimuli are emitted, thus making real rest possible.

Table 1.1 summarizes potential signs for inattention, impulsiveness, and hyperactivity.

1.2.2 Autism Spectrum (AS)

Autism is a processing difference that can have an impact on many areas of a person's life. Autistic people often experience differences in how they process information, their sensory environment, and how they interact with other people. People with autism spectrum (AS) disorders have to live with such constant stress. Employees with AS are real artists of life, who unfortunately our society labels as losers out of ignorance. But they are not losers, but amazing, valuable people who can be successful in life if we have people around us who recognize and, above all, acknowledge our strengths (Radley & Dart, 2023). But there are strategies that increase the chances of access to competitive employment in the open labor market.

Work and family are the building blocks of society in many countries and cultures. Both form networks of relationships and responsibilities that are often

Table 1.1 Signs of inattention, impulsiveness, and hyperactivity

Signs of inattention	Signs of impulsivity	Signs of hyperactivity
Often failing to pay close attention to details	Often using stronger and direct language	Often fidgeting and talking a lot to others
Having difficulties sustaining attention in work and play	Often interrupting or intruding on colleagues and team members	Difficulties sitting still in a group
Not seeming to listen when spoken to directly	Often having difficulties waiting to take turns	Often feeling restless when being with others
Often not following through on instructions and fails to finish tasks	Often reacting in direct and aggressive ways	Difficulties in waiting for others at work
Often having difficulty organizing tasks and activities	Being often on the go or acts as if "driven by a motor"	Often interrupting team members
Often avoiding, dislikes, or is reluctant to engage in tasks that require sustained mental effort	Often talking and arguing excessively with work colleagues and others	Often fidgeting with hands, feet or squirms
Often losing things. Easily being distracted by extraneous stimuli	Often blurting out answers before questions have been completed	Often leaving seat at workplace and elsewhere
Often being forgetful and not memorizing things of importance	Having difficulty playing or engaging in leisure activities quietly	Often running about or climbing excessively

Source: Author

complex and difficult even for people with good social skills. A workplace for employees with autism spectrum disorders should be designed in such a way that they can handle the tasks independently after a certain period of time. Their special interests, talents, or preferences can be of valuable service to society (Attwood & Garnett, 2024). It is crucial that the skills are used in the right place. The integration of people with AS into the world of work makes economic sense because they can make a very good contribution and relieves the burden on families, doctors, teachers, and social institutions.

For many employers, it would be worthwhile to fill the positions of the "baby boomers" who are retiring with employees with AS. However, most employers are unaware of the hidden talents in the autism spectrum population. But trained people with AS do good, often even excellent work in the right position and for the right employer. However, anyone who is honest in their application and admits that they are on the autism spectrum will be rejected in many cases. The person with AS hardly ever gets a chance, even though they will perform better in certain positions than a person with typical perception.

Leaders must take a gradual and sequential approach to achieve a mutual understanding of employees with the AS disorder. They must allow in their teams for a gradual approach to a mutual understanding of each other's view of the reality experienced together and the opportunity to learn from each other. Inclusive workforces and teams can experience the reality together and can take the chance to learn from each other (Kopp & Moser, 2020).

1.2.3 Tourette Syndrome (TS)

The Tourette syndrome (TS) is a neurological disorder that may cause sudden unwanted and uncontrolled rapid and repeated movements or vocal sounds called tics. TS is one of a group of disorders of the developing nervous system called tic disorders. Tourette syndrome or Tourette's syndrome is a common neurodevelopmental disorder that begins in childhood or adolescence. It is characterized by multiple movement tics and at least one vocal tic. Common tics are blinking, coughing, throat clearing, sniffing, and facial movements (Chaudhury, 2009). Tic is a neurological term. It refers to largely involuntary, sudden movements or the production of sounds or noises (Tourette Gesellschaft, 2024). It is important for you as a teacher to know what types of tics your student may exhibit. A distinction is made between motor and vocal tics.

Motor Tics (Movements)

- Face: Grimaces, blinking, rolling the eyes, head movements
- Shoulders and arms: Cramping of the fingers, shrugging of the shoulders, flinging the arms
- Trunk (less common): Dribbling with the feet, kicking

Special Motor Tics
- Touching: Touching other people, usually in the form of tapping
- Echopraxia: Imitating other people
- Copropraxia: Obscene gestures and actions

Vocal Tics (Noises)
- Sounds and noises: Coughing, clearing the throat, loud breathing in and out, squeaking, etc.
- Words and sentence

Special Vocal Tics
- Palilalia: Repetition of one's own words and sentences
- Echolalia: Reasonless repetition of what others say
- Coprolalia: Uttering obscene and socially unacceptable words and sentences

People with TS often share these concerns with everyone else in the workplace, but for their own special reasons. Although some people do suffer muscle pains, hoarseness or even injuries due to tics, TS is rarely "disabling" in the usual sense of the word. More often, it is social attitudes about physical and vocal tics that make trouble, and work is one place this trouble tends to boil over. Most employers still find accepting and accommodating wheelchair users and people with hearing impairments easier than understanding TS. Leaders must ensure that employees with the TS are accepted and recognized. Leaders can reduce stress factors. Relaxation techniques, biofeedback techniques and other behavioral therapy approaches can help to reduce stress reactions and improve self-control over tics. Dietary measures or depth psychological treatments alone have not proven to be sufficiently effective.

1.2.4 Mental Health Elements

Neurodivergent mental health disorders are expressed in stressed and anxient employees, who interact with the world and experience particular situations. Mental health symptoms are often supplemented by negative attitudes and a lack of understanding from others, which can lead to loneliness and feelings of isolation. The differences in understanding and relationships or social situations can distract people with these disorders. Mental health disorders can lead to challenges at work or in private life that may increase the risk of stress, anxiety, or depression.

Leaders should be talking about mental health and set time aside with no distractions. They should listen carefully to what employees tell them. Leaders can try to identify what the cause is, for example, by keeping questions open ended. By doing this, they can think about ways to help, for example, if they know about options for support at work or how to request reasonable adjustments.

1.2.5 Acquired Neurodiversity

Acquired neurodiversity refers to changes in a person's brain functioning following an injury, disease, or health condition. Rather than being a separate category, mental illnesses also come under the category of acquired neurodiversity. These may include disorders like depression, anxiety, and obsessive-compulsive disorder that can be episodic and have temporary effects on a person's cognition. Acquired Neurodiversity refers to neurodevelopmental conditions or differences acquired later in life due to external factors, such as brain injuries, illnesses, or trauma. Unlike innate neurodiversity, which is present from birth and typically refers to conditions like autism, ADHD, or dyslexia, acquired neurodiversity occurs due to events that happen after birth. These events can include traumatic brain injuries, strokes, infections, or neurodegenerative diseases. Acquired neurodiversity can manifest in various ways depending on the nature and severity of the underlying cause, and it may require different approaches to diagnosis, treatment, and support compared to innate neurodiversity.

Understanding and managing acquired neurodiversity is key to improving the quality of life for individuals affected. In this section, we will explore effective strategies and treatments that can be readily applied to support those with acquired neurodivergence:

Long-term prognosis varies, but with the right support and strategies, many individuals with acquired neurodiversity lead fulfilling, productive lives (Table 1.2).

1.2.6 Dyslexia

Dyslexia is a learning disorder that involves difficulty reading due to problems identifying speech sounds and learning how they relate to letters and words (decoding). Also called a reading disability, dyslexia is a result of individual differences in areas of the brain that process language. Dyslexia presents challenges with processing information, and these are exacerbated when people are stressed, so tasks become more difficult to

Table 1.2 Recommendations for acquired neurodiversity

Strategy	Description
Personalized plan	Tailoring the therapy in line with work routines to individual needs is crucial. Leaders must create an individual plan, which enables the employee to recover and rehabilitate
Supportive environment	Leaders must create a supportive work environment at home and work, with understanding from family and colleagues, and adjustments like reducing sensory overload
Routine and structure	Maintaining a consistent routine helps. A structured daily schedule can reduce anxiety
Technology aids	Utilizing technology like apps for communication or organization, to enhance independence and skills
Work team and community	Engaging with support groups for practical advice and daily routines can help the employee to recover and to re-enter the job

complete, it is more difficult to concentrate, and coping strategies become less effective. Employees with this disorder have secific and often outstanding skills such as big-picture thinking, lateral thinking and problem-solving, visual strengths and an intuitive understanding of how things work are often the hallmarks of successful dyslexic people.

Leaders and employers have a responsibility to make reasonable adjustments to the workplace to enable their dyslexic employees to carry out their roles to a satisfactory standard. Reasonable adjustments are changes to the working environment or way of working. Many adjustments are simple inexpensive changes that are easy to implement. They can be suggested (or requested) by either the employer or the member of staff. Leaders can give verbal as well as written instructions (this also applies to using voicemail rather than, or as well as, written memos) and provide assistive technology such as a screen-reader, scanning pen, text to speech, or mind-mapping software.

1.2.7 Dyscalculia

Dyscalculia is a learning difficulty that affects the employee's ability to master basic arithmetic, such as addition, subtraction, multiplication, and division. Math skills are essential in management and sales positions, cashier and clerical work, nursing, pharmacy and dentistry, and even jobs like landscaping, plumbing, electrical, and construction.

Adults with dyscalculia often take longer when working with numbers and are more prone to making errors in calculations. They may also experience higher levels of anxiety and frustration (Emerson & Babtie, 2014). Adults with dyscalculia may have a harder time learning and remembering math facts, such as multiplication tables. Estimation skills may also be affected. Dyscalculia is not a reflection of low intelligence, nor does it mean that an adult will not be successful at pursuing higher order mathematical thinking. Often, many people with dyscalculia believe they are simply bad at math. When dyscalculia is officially diagnosed, leaders can also give an adult access to appropriate accommodations, such as calculators for standardized tests. For others, a dyscalculia diagnosis represents a chance to learn new coping strategies for dealing with numbers that can lead to significant improvements in quality of life. There are apps that can help with daily finance management, automatically tracking expenses and making it easier to plan, schedule, and pay regular bills. Some adults with dyscalculia find that a diagnosis motivates them to take an online or in-person course to build their skills and prepare for a more fulfilling career.

1.2.8 Dyspraxia, Developmental Coordination Disorder (DCD)

Dyspraxia, also known as developmental coordination disorder (DCD), is a common disorder that affects movement and coordination. Dyspraxia does not affect your intelligence. It can affect your coordination skills—such as tasks requiring balance, playing sports, or learning to drive a car. The condition seems to disproportionately affect men and oftentimes runs in families. The term dyspraxia covers quite a broad range of deficiencies. In fact, as you should have known by the title, there are three types: motor dyspraxia, verbal dyspraxia, and oral dyspraxia.

Employees who are dyspraxic can also bring significant benefits to their workplaces. Dyspraxic workers are often hardworking, creative, and strong lateral thinkers. Most dyspraxic workers will have had to overcome significant challenges to be in the position they are now in. Leaders can embrace a work culture that takes the potential benefits from employees with DCD. Big picture thinking, problem-solving, tenacity, creativity, and empathy are all qualities associated with DCD.

1.3 Leadership as a Crucial Role for Embracing Neurodiversity

Neurodivergent employees are often subjected to bias when applying and interviewing for job positions in enterprises. Specifically, neurodivergent individuals can have their social engagement style compared to neurotypical individuals, which can affect their ability to obtain a job position. Stigmas against neurodivergence and cognition challenges in social situations can hinder an individual's ability to perform well in a traditional job interview. Organizations should aim to use neurodivergent employees' particular skills, such as pattern recognition, detection of deviations, attention to detail, analytical thinking, and extended focus, in the workforce, as well as educate companies on supporting neurodivergent employees.

Embracing neurodiversity within employees and leaders is a great way for organizations to grow, scale, and become top industry leaders. The companies that do, time and again, have excelled in productivity, innovation, risk reduction, retention, employee engagement, and revenue generation.

Multiple studies have highlighted that cognitive diversity in teams directly leads to innovation, creativity, and problem-solving abilities. A report on neurodiversity suggests that diverse teams outperform homogeneous ones in terms of profitability by more than 35% (Dennison, 2024). By taking advantage of neurodiversity and the unique skills it brings, companies can foster a culture of innovation, which improves teams' overall performances and competitiveness.

Figure 1.2 depicts 3C-leadership and neurodiversity matrix with recommendations for leaders.

1.4 Being Different, Making the Difference

Neurodivergent people make the difference by being different. They often know what their strengths and weaknesses are, which is built from a lifetime of creating workarounds to allow them to reach their full potential. Neurodiversity urges us to discuss brain diversity using the same kind of discourse that we would use when talking about biodiversity and cultural diversity. It can lead to the betterment of a company by hiring individuals who think differently and have different perspectives

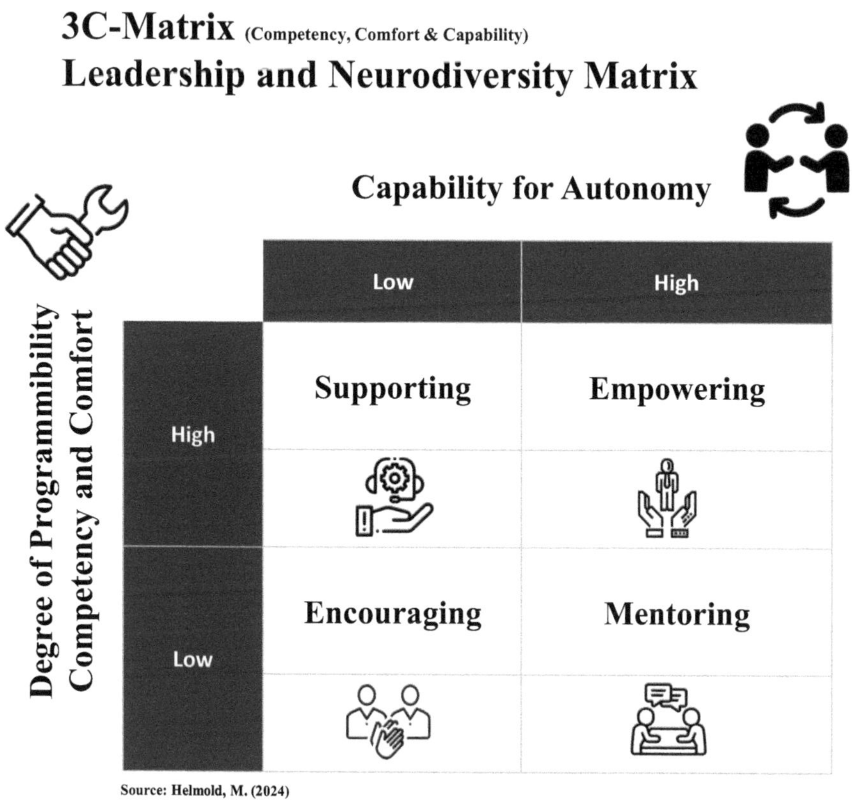

Fig. 1.2 3C-leadership and neurodiversity matrix

and opinions about problems and challenges at hand. Diversity in any sense should be seen as a strength and something that companies should try to implore (Austin & Pisano, 2017). The ability to think differently and think outside of the box is an extremely marketable skill for innovation and change.

Neurodiversity describes the variation of cognitive functions and behavioral traits in people. When it comes to the workplace, a neuro-inclusive perspective encourages the acceptance and recognition of neurodivergent staff—those individuals with varying neurotypes; recognizing and valuing the differences in how people's brains function, with unique ways of thinking, processing, and engaging. Neurodiversity, or neuro inclusion, challenges outdated perceptions that certain cognitive conditions are disorders needing to be fixed, instead celebrating those natural variations, considering them to be a normal part of human diversity. Neurodiversity in the workplace encourages employers to recognize and utilize the strengths of neurodivergent individuals, such as enhanced problem-solving skills, creativity, and attention to detail. Neurodiverse individuals have distinct spikes in their skillsets and certain "dips" compared to neurotypical workers, whose profiles are less varied. The peaks set neurodiverse talent apart. Neurodiverse individuals often excel in divergent thinking and creative problem-solving involving multiple solutions to a single problem. This ability to think "outside the box" can lead to groundbreaking innovations that might not be immediately apparent to neurotypical individuals.

1.5 Recommendations for Leaders in Embracing Neurodiversity

Authors and practitioners in leadership promote the leadership concept that neurodiverse-inclusive teams and leadership have better performance outcomes and higher productivity levels. To ensure that leadership teams are less homogenous but effectively accommodate neurodivergent individuals, companies must review current Human Resources (HR) policies and practices and focus on culture and environment to avoid tokenistic actions that, in the long run, will be harmful to neurodivergent people. Nurturing neurodiversity in leadership is essential for companies to excel and maintain a competitive edge (Patten et al., 2024). Creating an inclusive leadership team is not just about fulfilling a social responsibility but also about harnessing untapped potential that can drive innovation and growth (Helmold, 2022). To effectively build a neurodiverse-inclusive leadership team, the following managerial recommendations are crucial for performance excellence:

- Leaders and organization must invest in neurodiversity awareness and leadership development to retain neurodiverse high performers, utilizing various formal and informal professional development tools, such as mentoring, coaching, and education opportunities.
- Leaders must ensure that neurodiverse employees are included in the career programs and leadership development pipeline through regular talent evaluations.
- Leaders must incorporate neurodiverse representation by ensuring that the leadership team has neurodiverse members, either by actively seeking out neurodiverse candidates or by providing opportunities for existing employees to develop their leadership skills.
- Leaders should partner with neurodiversity organizations if possible since most organizations are not experts in the field of neurodiversity.
- Leaders must prioritize systemic inclusion by removing barriers to entry, such as job descriptions and interviewing processes, and ensuring equal access to opportunities for growth and leadership, organizations can fully harness the unique strengths and abilities of neurodivergent talents.
- Leaders should provide pathways for career development, since neurodivergent employees face structural barriers to career development owing to ingrained leadership stereotypes and assumptions about their ambitions.
- Leaders should collaborate with neurodivergent employees to identify performance goals, skill development interests, and tailored growth plans.

1.6 Case Study: EU Commission Embracing Neurodiversity

The European Commission is fostering the embracement of neurodiversity and neurodivergence in organizations throughout the European Union. The aim of the research proposal "Neurodiversity between Law and Science" (NEDBELS) is to inquire into the legal impacts and sociopolitical implications of the concept of neurodiversity. This term pertains to individuals diagnosed with neurodevelopmental disorders such as Asperger's syndrome and Autism and hypothesizes the emergence of a new category of difference in the human population (EU Commission, 2024).

NEDBELS is examining how the concept challenges the principle of equality, and its impact on legal systems and practices. The project was launched to address a perceived gap in communications between legal experts and scientists on the treatment of those with neurodevelopmental disorders—which affects formal training and daily practice in diagnosis, education, and health services for those affected (EU Commission, 2024).

NEDBELS explores how this concept challenges the constitutional principle of equality, as well as how it fosters the need to accommodate new principles in criminal and civil law. This proposal has risen from a need for heightened communication between law and science. Scientific theories, concepts, and methods widely permeate law and adjudication, and legal and ethical frameworks intimately drive applied sciences, as in education and healthcare systems. Therefore, one might reasonably expect that these two worlds would cooperate on a regular basis. However, lawyers, scientists, and healthcare professionals operate largely insulated from one another in research, formal training, and daily practice. This divergence prevents vulnerable minorities, such as people with neurodevelopmental disorders, from reaping the full sociopolitical potential of scientific advancements. NEDBELS's goal is to increase communication between law and neurosciences through interdisciplinary research methodologies and scientific training-through-research as well as by enhancing the synergy among legal experts, policymakers, and scientists in the field of neurodiversity (EU Commission, 2024).

References

Amstrong, T. (2011). *The power of neurodiversity. Unleashing the advantages of your differently wired brain*. Hachette Books.

Attwood, T., & Garnett, M. (2024). Mit Asperger-Syndrom im Job zufrieden und erfolgreich sein: Das 7-Schritte-Programm für Menschen im Autismus-Spektrum. Trias Open Library.

Austin, R. D., & Pisano, G. P. (2017). Neurodiversity as a competitive advantage—why you should embrace it in your workforce. Harvard Business Review. Retrieved February 1, 2025. https://hbr.org/2017/05/neurodiversity-as-a-competitive-advantage.

Blume, H. (1998). Neurodiversity. On the neurological underpinnings of geekdom. Retrieved December 28, 2024. https://www.theatlantic.com/magazine/archive/1998/09/neurodiversity/305909/.

Chaudhury, U. (2009). Tics und Tourette-Syndrom: Ein Handbuch für Fachleute und Eltern. DGVT Verlag Frankfurt.

Dennison, K. (2024). The benefits of neurodiversity in leadership. Forbes. Retrieved December 24, 2024. https://www.forbes.com/sites/karadennison/2024/11/14/the-benefits-of-neurodiversity-in-leadership/.

Emerson, J., & Babtie, P. (2014). *The dyscalculia solution: A step-by-step guide to teaching children with numeracy difficulties*. Bloomsbury Publishing.

EU Commission (2024). Horizon 2020. NEURODIVERSITY BETWEEN LAW AND SCIENCE Retrieved January 5, .2025. https://cordis.europa.eu/project/id/656988.

Helmold, M. (2022). Leadership. Agile, virtuelle und globale Führungskonzepte in Zeiten von neuen Megatrends. Springer Wiesbaden.

Kapp, S. K. (2020). *Autistic community and the neurodiversity movement. Stories from the frontline*. Springer.

Kopp, G., & Moser, K. (2020). *Eine Welt—zwei Wahrnehmungen. . Wie Autisten die Welt erleben und was sie darüber denken*. Springer.

Lachenmaier, H. (2023). ADHD and success at work. How to turn supposed shortcomings into strengths. Springer .

Marks, D. (1999). *Disability: Controversial debates and psychosocial perspectives*. Psychology Press.

Patten, E., et al. (2024). *Neurodiversity and work. Employment, identity, and support networks for neurominorities*. Springer.

Radley, K.C. and Dart, E.H. (2023). Vermittlung sozialer Kompetenzen für Menschen mit Autismus. Integration der Forschung in die Praxis. Springer .

Tourette Gesellschaft. (2024). Studien. Retrieved December 24, 2024. https://tourette-gesellschaft.de/.

2 Neurodiversity, Talent Shortage, and Stress: Rationale and Call for Action for Leaders

2.1 Understanding Neurodiversity in the Workforce

The modern workforce is grappling with several critical issues, and leaders have a significant role to play in addressing them. One major area of focus is neurodiversity, which refers to the natural differences in how people think, learn, and solve problems. This concept is transforming how organizations think about talent by highlighting the unique strengths individuals with conditions like autism, ADHD, and dyslexia can bring to the workplace. These strengths, however, often require tailored support systems to fully flourish (Austin & Pisano, 2017). Alongside this, organizations are struggling with a global talent shortage, as the demand for skilled workers continues to outpace supply. This gap in the labor market is compounded by increasing levels of workplace stress, which negatively affect productivity, employee well-being, and organizational resilience (World Health Organization, 2022). This chapter explores these three intertwined challenges and provides practical strategies for leaders to create inclusive workplaces where all employees can thrive (Fig. 2.1).

2.1.1 Foundations

Neurodiversity refers to the idea that people's brains function in diverse ways, leading to unique approaches to learning, problem-solving, and thinking. Historically, conditions like autism and ADHD were seen primarily as obstacles. However, today, these differences are being recognized for their contributions to creativity, innovation, and problem-solving (Silberman, 2015). For instance, individuals with autism often excel at recognizing patterns and paying close attention to detail, while those with ADHD bring creativity and rapid decision-making skills to the workplace. The topic of neurodiversity can be examined from three key perspectives: social, ethical, and economic. Socially, it

M. Helmold, M. Martensen, *Leadership in Neurodiversity and Neurodivergence*, Diversity and Inclusion Research, https://doi.org/10.1007/978-3-032-09343-1_2

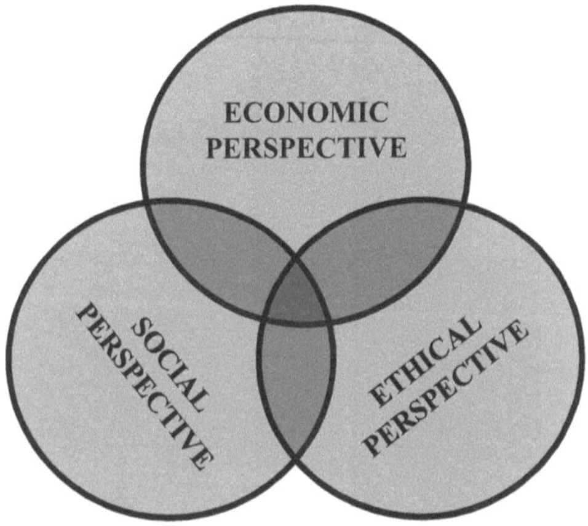

Fig. 2.1 Perspectives on neurodiversity

emphasizes the importance of valuing diverse abilities within communities. Ethically, it highlights the need to provide equal opportunities for all individuals. Economically, it showcases the untapped potential of neurodiverse talent to drive innovation and growth and thus help overcome talent shortage. This growing appreciation for neurodiversity has prompted organizations to rethink job roles and workflows, tailoring them to fully leverage the strengths of neurodivergent employees. By doing so, companies not only support individual workers but also enhance overall team performance, fostering complementary skill sets and diverse perspectives.

2.1.2 Key Challenges

Despite these strengths, neurodivergent individuals face considerable barriers in the workplace. Recruitment processes often unintentionally favor neurotypical candidates. For example, traditional interviews typically prioritize social skills and verbal communication, which can disadvantage individuals with autism (Rollnik-Sadowska & Grabinska, 2024). Additionally, workplace environments are usually designed with neurotypical needs in mind. Features such as open office layouts, unclear communication practices, and rigid work schedules can create unnecessary stress for neurodivergent employees. These challenges make it difficult for them to thrive and fully contribute their talents. According to the National Autistic Society (2021) and the Office for National Statics (2021), only 22% of autistic adults in the UK hold full-time employment, underscoring the systemic gaps that need to be addressed. To unlock the full potential of this workforce, organizations must implement systemic changes that foster inclusion and accessibility.

2.1.3 Opportunities for Organizations

Organizations that embrace neurodiversity stand to gain significant benefits. For instance, programs like SAP's Autism at Work have shown that neurodiverse teams can excel in tasks requiring precision, focus, and innovative thinking (Austin &

Pisano, 2017). Neurodivergent employees often approach problems from unique perspectives, leading to creative solutions and improved processes. Beyond operational performance, hiring neurodivergent individuals fosters a more inclusive workplace culture, where diverse abilities are valued and celebrated. This inclusivity not only improves team morale and retention but also enhances the organization's reputation. Companies that prioritize neurodiversity often attract socially conscious clients, partners, and investors who value equity and inclusion. By creating environments where all employees can thrive, organizations position themselves for long-term success while making meaningful strides toward social responsibility.

2.2 Talent Shortage: A Catalyst for Change

The global talent shortage presents a significant challenge, with an estimated gap of 85 million workers projected by 2030 (ManpowerGroup, 2022). This shortage stems from several factors, including demographic shifts such as aging populations, declining birth rates in many countries, and rapid technological advancements that demand new and evolving skill sets. The growing reliance on digital transformation and automation has particularly heightened the demand for specialized technical skills, creating an urgent need for qualified workers. Unfortunately, education systems and traditional hiring methods have struggled to keep pace with these changes. Many companies find themselves ill-equipped to bridge this gap, leaving critical roles unfilled and creating significant operational challenges (Fig. 2.2).

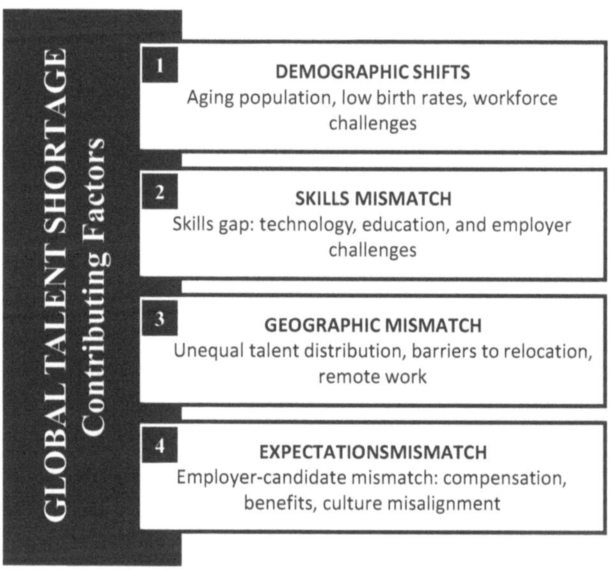

Fig. 2.2 Global talent shortage: contributing factors (Manatal, 2024)

The global talent shortage is a significant challenge driven by multiple factors that require strategic solutions. These factors can be categorized into four main dimensions: demographic shifts, skills mismatch, geographic mismatch, and expectations mismatch.

2.2.1 Demographic Shifts

Demographic changes, such as aging populations and declining birth rates in developed countries, have significantly reduced the size of the active workforce. This phenomenon is further exacerbated by longer life expectancies and delayed retirement, as older workers remain in the workforce longer, limiting opportunities for younger generations to advance. Additionally, certain industries experience greater strain as the aging population demands more healthcare and social support services, compounding workforce shortages in these sectors. The uneven distribution of employment opportunities across age groups highlights a growing disparity in labor markets and underscores the importance of addressing these demographic challenges (Manatal, 2024).

2.2.2 Skills Mismatch

A major contributor to the talent shortage is the skills mismatch, where the qualifications and skills employers seek are not aligned with the capabilities of the available workforce. The rapid pace of technological advancements in fields such as artificial intelligence, data science, and programming has outstripped the ability of traditional education systems to adapt. As a result, many graduates enter the workforce with theoretical knowledge but lack the practical skills needed to succeed in today's competitive labor market. Additionally, industries undergoing digital transformation demand niche expertise that is often not covered by conventional training programs. Employers also struggle to invest in and design effective upskilling initiatives that keep pace with the shifting demands of the modern economy, further widening the skills gap (Manatal, 2024).

2.2.3 Geographic Mismatch

Geographic mismatch refers to the uneven distribution of talent and job opportunities across regions. While urban centers often have an oversupply of skilled workers, rural or economically disadvantaged areas face acute shortages. Barriers such as high relocation costs, restrictive immigration policies, and regional disparities in housing affordability make it challenging to address these geographic imbalances. Even with the rise of remote work, discrepancies in access to reliable technology and digital infrastructure have limited the ability of certain regions to bridge the gap. This mismatch highlights the need for targeted policy interventions and investment in infrastructure to ensure more equitable talent distribution.

2.2.4 Expectations Mismatch

Expectations mismatch arises when there are discrepancies between what candidates seek and what employers are willing to offer. For example, candidates increasingly value flexibility in work arrangements, such as hybrid or fully remote roles, while some employers continue to emphasize traditional in-office setups. Salary expectations, benefits packages, and workplace culture are other areas where mismatches frequently occur. Organizations that fail to adapt to these evolving preferences risk losing access to top talent. To address this, companies need to reevaluate their employee value propositions and align their offerings with the current expectations of job seekers in a competitive labor market.

2.2.5 The Economic Impact

The consequences of the global talent shortage extend far beyond unfilled roles. According to projections, unaddressed talent gaps could result in $8.5 trillion in lost global economic output by 2030 (Korn Ferry, 2021). This economic loss stems from reduced productivity, slower innovation cycles, and higher costs associated with recruitment and training. Furthermore, organizations that fail to fill critical positions face diminished operational efficiency and growing strain on existing employees, leading to increased burnout and turnover rates. These challenges emphasize the urgent need for businesses to develop proactive strategies that mitigate the economic fallout of the talent shortage.

2.2.6 Neurodiversity as a Solution

Integrating neurodivergent individuals into the workforce offers a practical and innovative solution to the talent shortage. Neurodivergent individuals bring unique perspectives and specialized skills that align closely with many in-demand roles. For instance, individuals with dyslexia often excel in visual and creative fields, contributing fresh ideas to industries like advertising, design, and media. Similarly, those with ADHD thrive in dynamic, fast-paced environments where problem-solving, multitasking, and adaptability are critical. By focusing on the strengths and competencies of neurodivergent candidates rather than traditional qualifications, employers can tap into a vast, underutilized talent pool. In addition to addressing skill gaps, promoting neurodiversity enhances workplace innovation and strengthens team dynamics. Companies like SAP and Microsoft have demonstrated that neurodiverse hiring initiatives lead to measurable improvements in performance, culture, and reputation. Moreover, these programs align with broader social and ethical goals, showcasing a commitment to inclusivity and corporate social responsibility. By embracing neurodiversity, organizations not only mitigate workforce shortages but also build more dynamic, forward-thinking, and resilient teams (Table 2.1).

Table 2.1 The benefits of neurodiverse teams (adapted from Austin and Pisano, 2017)

Strength	Neurodivergent condition	Example applications
Pattern recognition	Autism	Data analysis, cybersecurity
Creative problem-solving	ADHD	Advertising, design
Attention to detail	Autism, dyslexia	Quality assurance, research

2.3 Workplace Stress: The Silent Epidemic

2.3.1 Causes and Impacts of Workplace Stress

Workplace stress has become a pervasive issue in modern organizations, earning its title as the "silent epidemic." It affects employees across industries and roles, manifesting in various forms such as physical health problems, mental exhaustion, and decreased productivity. According to the American Psychological Association (2021), workplace stress contributes to over $300 billion in annual costs for U.S. businesses due to absenteeism, turnover, and reduced performance. Understanding the nature and causes of workplace stress is essential for leaders aiming to create healthier and more productive work environments.

Stress arises from numerous sources, with excessive workloads, lack of control, and sensory overload being particularly significant contributors. Excessive workloads are frequently cited by employees as a major cause of anxiety and burnout, especially when expectations are poorly managed or resources are insufficient to meet demands. This challenge is even more pronounced for neurodivergent individuals who may process tasks differently or require additional time to ensure accuracy. Overburdening such employees can lead to chronic stress and disengagement, undermining their unique strengths and contributions. Lack of job control further exacerbates stress. Employees who feel powerless to influence decisions that affect their work often experience frustration and disillusionment. For neurodivergent individuals, this lack of autonomy can be particularly detrimental, as they may thrive in structured environments with clear guidelines. Ambiguity and shifting expectations only intensify the pressures they face, leaving them vulnerable to feelings of inadequacy.

2.3.2 Addressing Stress Through Leadership and Support

Sensory overload is a critical issue for neurodivergent employees. Open-plan offices, a staple of modern workplace design, often fail to accommodate those with sensory sensitivities. The constant noise, visual distractions, and interruptions can create an environment that is not only unproductive but also harmful to mental well-being. Addressing these stressors requires thoughtful accommodations, such as providing access to quiet workspaces or allowing remote work options to minimize unnecessary exposure to sensory triggers.

The effects of workplace stress extend beyond the individual, impacting organizational performance and culture. Employees under chronic stress are more likely to take sick days, deliver suboptimal work, or leave their roles altogether. High turnover rates and absenteeism are costly for organizations, not just financially but also in terms of lost institutional knowledge and morale (Demerouti et al., 2001). Moreover, stress undermines creativity and collaboration, stifling the innovation that is critical for businesses to thrive in competitive markets.

Leaders play a pivotal role in addressing workplace stress. By fostering an environment that prioritizes mental well-being, they can mitigate the negative consequences of stress and create a more resilient workforce. This involves not only implementing accommodations for neurodivergent employees but also cultivating a broader culture of support and inclusion. For example, offering flexible work hours and clear communication channels can help employees manage their responsibilities more effectively. Leadership training programs that emphasize empathy, awareness of neurodiversity, and stress management equip managers with the tools to identify and address stressors proactively.

Measuring and monitoring stress levels are essential for developing effective interventions. Tools such as employee surveys, focus groups, and regular check-ins provide valuable insights into the stressors affecting the workforce. Armed with this information, organizations can implement targeted strategies to reduce stress and promote well-being. Initiatives such as mindfulness training, access to mental health resources, and opportunities for professional development empower employees to navigate workplace challenges more effectively. Ultimately, tackling workplace stress requires a holistic approach that recognizes the diverse needs of employees. For neurodivergent individuals, this means moving beyond generic solutions to provide tailored support that enables them to thrive. By addressing the root causes of stress and fostering a culture of inclusion, leaders can transform the workplace into a space where all employees feel valued and supported.

2.4 Best Practices for Inclusive Leadership

2.4.1 Talent Acquisition

Inclusive recruitment practices are the foundation of diverse workplaces. Organizations must rethink traditional hiring methods to eliminate bias and create equitable opportunities for all applicants, including those who are neurodivergent. For example, replacing subjective interview evaluations with skill-based assessments allows candidates to showcase their capabilities without being unfairly judged on social skills, which may not align with their strengths (Austin & Pisano, 2017). Likewise, a holistic employee value proposition emphasizing inclusivity and personal growth sends a clear message to prospective hires that the organization prioritizes diversity and employee well-being. Accessibility must also underpin the recruitment process. By designing barrier-free candidate experiences, organizations ensure that everyone—regardless of their unique needs—has a fair opportunity to succeed. Practical measures include clear communication about interview expectations, the availability of virtual assessments, and flexible scheduling to accommodate neurodivergent candidates. These practices not only enhance fairness but may also help companies attract exceptional talent often overlooked in traditional hiring models.

2.4.2 Talent Development

Developing talent is equally critical to fostering a truly inclusive workplace. Training programs tailored to neurodiversity should aim to build awareness among all employees, equipping them with the skills to collaborate effectively in diverse teams. Leadership development initiatives are especially important, as they shape the organizational culture from the top down. Programs that address stress prevention and unconscious bias, foster empathy, and promote effective communication can empower managers to create supportive environments for neurodivergent team members. Furthermore, tailored development pathways for neurodivergent employees not only enhance retention but also encourage individual growth. Providing access to mentors, professional development opportunities, and specialized training fosters a sense of belonging and empowerment. Employees who feel supported in their career journeys are more likely to remain engaged and committed to organizational goals.

2.4.3 Talent Retention and Motivation

Retaining talent involves creating a workplace culture that values individuality and prioritizes inclusion. Leadership models must emphasize trust, empathy, and collaboration. Transformational leadership, for example, has been shown to drive engagement and motivation by aligning organizational values with employees' personal goals (Goleman, 2000). An inclusive culture, where employees feel respected and valued, reduces turnover and strengthens team cohesion. Motivation also stems from recognition. Leaders can sustain employee morale by acknowledging unique contributions and providing regular, constructive feedback. Recognition programs designed to accommodate diverse preferences—such as public acknowledgments, personalized rewards, or opportunities for skill building—foster a positive and motivated workforce.

2.4.4 Working Conditions

The physical and structural design of the workplace significantly impacts employee well-being. Flexible working hours, remote options, and adaptable schedules allow employees to balance personal and professional demands effectively and may help reduce stress levels. Neurodivergent employees often benefit from such flexibility, as it reduces environmental stressors and enables them to perform at their best. Studies show that organizations offering flexibility report higher levels of employee satisfaction and productivity. Promoting eustress, or positive stress, is another vital component of effective workplace management. Eustress encourages growth and innovation by presenting employees with meaningful challenges, but without the detrimental effects of distress. Policies that offer clear guidance, ample support, and resources to handle challenges ensure that employees experience stress in a way that drives, rather than hinders, their development.

2.4.5 Diverse Talent Pools

Finally, organizations must tap into diverse talent pools by valuing individuality and recognizing the potential of neurodivergent individuals. Leveraging the unique strengths of a neurodiverse workforce fosters creativity and innovation. For instance, individuals with autism spectrum disorder (ASD) often excel in roles requiring attention to detail and pattern recognition, while those with ADHD may bring creativity and high-energy problem-solving skills to the table (Hendricks, 2010). Encouraging individuality within a structured, supportive framework ensures that employees can achieve their full potential and contribute meaningfully to the organization's success.

Neurodiversity, talent shortages, and workplace stress are deeply interconnected challenges that require new approaches from leaders. By adopting inclusive practices, addressing employee well-being, and focusing on skills-based hiring, organizations can unlock the potential of neurodiverse talent and build stronger, more resilient teams. Microsoft's example shows that with the right strategies, organizations can thrive while making the workplace better for everyone (Fig. 2.3).

2.5 Case Study: Embracing Neurodiversity at Microsoft

Microsoft, one of the world's largest technology companies, recognized the immense value of tapping into neurodiverse talent pools through its groundbreaking Neurodiversity Hiring Program. Launched in 2015, this initiative was designed to address the challenges neurodivergent individuals face during traditional recruitment processes and create an environment where their unique skills could thrive (Hennekam & Follmer, 2024; Microsoft, 2025; Mentra, 2024; Thomas, 2023).

2.5.1 Recruitment and Onboarding

Microsoft reimagined its recruitment process to make it more inclusive for neurodivergent candidates. Instead of relying on conventional interviews, which often place undue emphasis on social interactions, the company implemented a multi-day evaluation process. This approach allowed candidates to demonstrate their technical

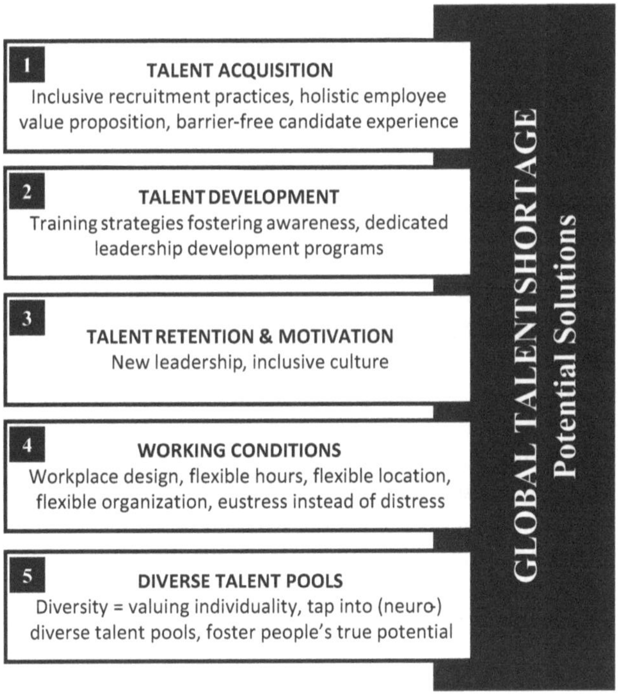

Fig. 2.3 Global talent shortage: potential solutions (author's own figure)

skills and problem-solving abilities through practical exercises. For example, candidates for software engineering roles participated in collaborative coding projects, enabling them to showcase their strengths in a supportive, low-pressure setting. To attract neurodiverse talent, Microsoft partnered with organizations like Specialisterne and Autism Speaks, which specialize in connecting neurodivergent individuals with meaningful employment opportunities. The company also ensured transparency in its job postings and provided detailed information about the interview process to reduce candidate anxiety.

2.5.2 Workplace Adjustments and Support

Once candidates joined Microsoft, the company provided tailored support to help them succeed. Microsoft also trained managers to better understand the needs of neurodivergent employees, fostering a culture of empathy and inclusion. The company emphasized mentorship as a cornerstone of its program. New hires were paired with mentors who provided guidance on workplace expectations and career development. Regular one-on-one check-ins allowed mentors and mentees to address any challenges early and collaboratively develop strategies for success.

2.5.3 Outcomes and Lessons Learned

Strategic initiatives such as Microsoft's Neurodiversity Hiring Program can yield significant benefits, both for the organization and for its employees. Neurodivergent team members can make substantial contributions in areas such as software testing, cybersecurity, and data analysis (Austin & Pisano, 2017). For example, employees on the autism spectrum can use their exceptional attention to detail to identify critical vulnerabilities in a software system that would potentially be overlooked by others. Such programs can also have a positive impact on workplace culture. Neurotypical employees often report increased awareness and appreciation for diversity, and teams may become more innovative by incorporating a wider range of perspectives (Hennekam & Follmer, 2024). Hence, a program such as Microsoft's Neurodiversity Hiring Program not only addresses talent shortages but also enhances overall team performance and cohesion. Microsoft's initiative has inspired other organizations to adopt similar programs, demonstrating the transformative potential of embracing neurodiversity (Hennekam & Follmer, 2024). By focusing on inclusive recruitment, providing tailored accommodations, and fostering a supportive environment, Microsoft has shown that investing in neurodiverse talent is not only a moral imperative but also a strategic advantage.

2.5.4 Conclusion

Microsoft's Neurodiversity Hiring Program highlights the value of intentional efforts to include neurodivergent individuals in the workforce. By addressing barriers in recruitment and creating an inclusive workplace, Microsoft has unlocked the potential of underutilized talent pools, driving innovation and setting a precedent for other organizations to follow. Their success underscores the importance of leadership commitment in creating workplaces where everyone can thrive.

References

American Psychological Association. (2021). *Stress in America 2021*. APA.

Austin, R. D., & Pisano, G. P. (2017). Neurodiversity as a competitive advantage. *Harvard Business Review*.

Demerouti, E., Bakker, A. B., Nachreiner, F., & Schaufeli, W. B. (2001). The job demands-resources model of burnout. *Journal of Applied Psychology, 86*(3), 499–512.

Goleman, D. (2000). Leadership that gets results. *Harvard Business Review*.

Hendricks, D. (2010). Employment and adults with autism spectrum disorders: Challenges and strategies for success. *Journal of Vocational Rehabilitation, 32*(2), 125–134.

Hennekam, S., & Follmer, K. (2024). Neurodiversity and HRM: a practice-based review and research agenda. *Equality, Diversity and Inclusion, 43*(7), 1119–1129.

Korn Ferry. (2021). The global talent crunch. Retrieved January 28, 2025. https://www.kornferry.com.

Manatal. (2024). 5 Best strategies to mitigate talent shortage. Retrieved February 3, 2025. https://www.manatal.com/blog/talent-shortage.

ManpowerGroup. (2022). Talent shortage 2022: Closing the skills gap.

Mentra. (2024). Why every neurodivergent should want to work at Microsoft: A leader in inclusive hiring. Retrieved February 3, 2025. https://www.mentra.com/post/why-every-neurodivergent-should-want-to-work-at-microsoft-a-leader-in-inclusive-hiring.

Microsoft. (2025). Microsoft neurodiversity hiring program & FAQ. Retrieved February 3, 2025. https://www.microsoft.com/en-us/diversity/inside-microsoft/cross-disability/neurodiversityhiring?utm_source=chatgpt.com&oneroute=true.

National Autistic Society. (2021). Autistic adults and employment: Current challenges and opportunities. Retrieved January 23, 2025. https://www.autism.org.uk.

Office for National Statistics. (2021). Outcomes for disabled people in the UK: 2020. Retrieved January 3, 2025. https://www.ons.gov.uk/peoplepopulationandcommunity/healthandsocialcare/disability/articles/outcomesfordisabledpeopleintheuk/2020.

Rollnik-Sadowska, E., & Grabinska, V. (2024). Managing neurodiversity in workplaces: A review and future research agenda for sustainable human resource management. *Sustainability, 16*, 6594.

Silberman, S. (2015). *NeuroTribes: The legacy of autism and the future of neurodiversity*. Avery.

Thomas, S. (2023). Microsoft opens 'different front door' to neurodiversity. Retrieved February 3, 2025. https://www.hcamag.com/us/specialization/diversity-inclusion/microsoft-opens-different-front-door-to-neurodiversity/464063.

World Health Organization. (2022). World mental health report: Transforming mental health for all.

Recruitment and Onboarding of Neurodivergent Employees

3

3.1 Introduction

The workplace is increasingly recognized as a space where diversity is not only embraced but also strategically utilized to drive innovation, creativity, and productivity. Among the various dimensions of diversity, neurodiversity—referring to the natural variation in neurological functioning among individuals—has gained significant attention. The term, first introduced by sociologist Judy Singer in the late 1990s, encompasses individuals with conditions such as autism, attention deficit hyperactivity disorder (ADHD), dyslexia, and other cognitive differences (Singer, 2016). These conditions, often considered disabilities in traditional contexts, are now viewed as valuable assets in workplaces that foster inclusion.

Despite its promise, the inclusion of neurodivergent individuals in the workplace remains fraught with challenges. Conventional recruitment and onboarding practices often fail to accommodate the unique needs and strengths of neurodivergent employees, leading to high barriers to entry. Traditional interviews, for instance, place undue emphasis on verbal communication and social interaction, areas where some neurodivergent individuals may experience difficulties (Austin & Pisano, 2017). Such practices inadvertently exclude a significant portion of the talent pool, reinforcing systemic inequities and depriving organizations of the innovative potential that neurodivergent individuals bring.

Neurodiversity, however, is not merely a matter of social responsibility—it is also a strategic imperative. Organizations that embrace neurodivergent talent often report enhanced problem-solving capabilities, improved attention to detail, and the generation of creative solutions to complex challenges. For example, autistic employees often excel in roles requiring pattern recognition and precision, while individuals with ADHD bring energy, creativity, and adaptability to dynamic environments (Silberman, 2015). Leveraging these strengths requires rethinking

M. Helmold, M. Martensen, *Leadership in Neurodiversity and Neurodivergence*, Diversity and Inclusion Research, https://doi.org/10.1007/978-3-032-09343-1_3

traditional human resource practices to ensure a level playing field for neurodivergent candidates.

The importance of inclusive recruitment and onboarding practices extends beyond the ethical and economic dimensions; it aligns with broader organizational goals of innovation and resilience. Research indicates that diverse teams are more effective at solving complex problems and are up to 35% more likely to outperform less diverse teams in terms of profitability (Hewlett et al., 2013). Furthermore, as global talent shortages intensify, neurodivergent individuals represent an untapped talent pool that can help bridge critical skills gaps, particularly in fields like technology, research, and data analytics (Austin & Pisano, 2017).

This chapter examines the challenges and opportunities associated with recruiting and onboarding neurodivergent employees. By addressing systemic barriers and promoting inclusive practices, organizations cannot only fulfill their ethical responsibilities but also unlock the potential of an underutilized workforce. In doing so, they can foster a workplace culture that celebrates diversity in all its forms, ensuring that all employees—neurodivergent and neurotypical alike—are empowered to thrive. In subsequent sections, we will delve into the unique characteristics of neurodivergent talent, explore strategies to redesign recruitment processes, and propose effective onboarding practices tailored to neurodivergent employees.

3.2 Understanding Neurodiverse Talent

Neurodivergent individuals—those with cognitive differences such as autism, ADHD, dyslexia, and others—bring unique strengths to the workplace that often remain underutilized. These strengths include advanced pattern recognition, exceptional problem-solving skills, and innovative thinking, all of which can drive organizational success. For example, individuals with autism frequently excel in detail-oriented tasks and data analysis, where precision is paramount, while those with ADHD demonstrate exceptional creativity and thrive in dynamic, fast-paced environments that require adaptability and quick decision-making (Silberman, 2015). Dyslexic employees often possess strong spatial reasoning and outside-the-box thinking, making them invaluable in design, architecture, and engineering. Beyond technical skills, many neurodivergent individuals bring fresh perspectives to problem-solving, challenging conventional approaches and driving innovation.

Despite these advantages, neurodivergent individuals face significant barriers during the recruitment process. Traditional hiring practices, such as unstructured interviews and vague job descriptions, often disadvantage these candidates by prioritizing social and verbal communication skills over technical abilities or problem-solving aptitude. Research indicates that over 75% of neurodivergent individuals feel misunderstood or undervalued during job interviews (Hendrickx, 2010). For instance, an unstructured interview may place undue weight on conversational fluency, inadvertently penalizing candidates who excel in analytical tasks but struggle with social interactions. This systemic bias excludes talented individuals from the workforce, limiting both their opportunities and the organization's ability to innovate.

Inclusive hiring practices can help bridge this gap by shifting the focus from cultural conformity to capability. Structured assessments and task-based evaluations are particularly effective, as they allow neurodivergent candidates to demonstrate their strengths in practical, low-pressure settings. For example, coding tasks, data analysis challenges, or design assignments provide an objective way to assess competencies directly relevant to the role. Similarly, providing clear, straightforward job descriptions that avoid jargon and ambiguity can reduce anxiety for candidates and make the hiring process more accessible. Organizations that implement such measures often see immediate benefits, including increased innovation, improved team dynamics, and higher retention rates (Hewlett et al., 2013).

To illustrate the potential of neurodivergent talent, organizations like SAP and Microsoft have implemented dedicated hiring programs designed to attract and support neurodivergent employees. SAP's Autism at Work initiative, for example, has shown that teams excel in roles requiring precision and innovative thinking, such as software testing, cybersecurity, and quality assurance (Austin & Pisano, 2017). Similarly, Microsoft's Neurodiversity Hiring Program focuses on creating supportive environments where neurodivergent employees can thrive. These programs not only address recruitment barriers but also foster long-term success through tailored onboarding processes, mentorship, and workplace accommodations. By integrating neurodivergent employees into key roles, these companies demonstrate how targeted recruitment and supportive environments can unlock the full potential of neurodivergent employees while creating a competitive advantage.

3.3 Rethinking Recruitment Processes

Traditional recruitment processes often serve as barriers rather than as gateways for neurodivergent candidates focusing on predicting performance in terms of right and wrong (see Fig. 3.1). These systems are largely structured to prioritize neurotypical traits, such as strong verbal communication skills, charisma during interviews, and adherence to standardized workflows. Consequently, neurodivergent individuals—whose strengths often lie in technical, analytical, or creative domains—may be overlooked, perpetuating systemic biases in hiring practices (Hennekam & Follmer, 2024). This exclusion has far-reaching consequences, depriving organizations of diverse cognitive approaches that can enhance problem-solving and innovation.

To bridge this gap, organizations must rethink recruitment as a holistic, inclusive process that focuses on assessing actual skills and potential rather than conformity to conventional standards. This transformation demands structural changes, including the adoption of alternative assessment tools, inclusive interview practices, and mechanisms to counteract bias at every stage of the hiring pipeline. By committing to these changes, companies can cultivate a workforce that reflects diverse perspectives and drives strategic success.

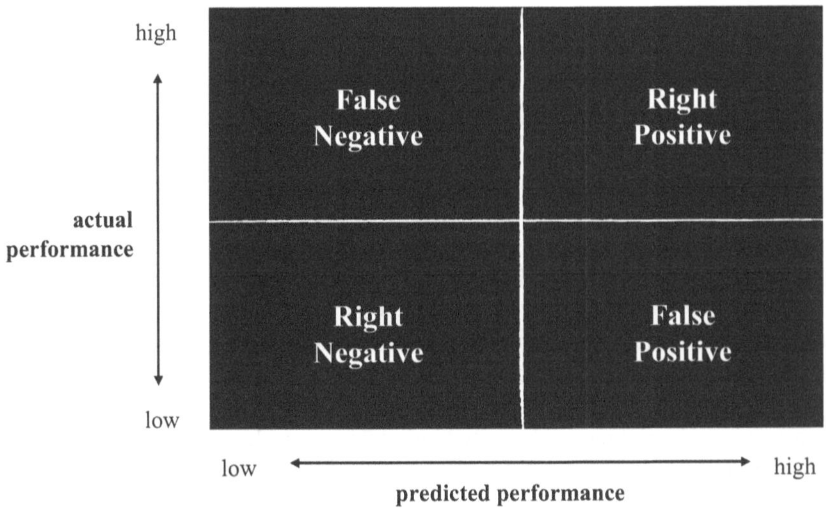

Fig. 3.1 Traditional recruitment paradigm

One key strategy involves moving away from reliance on traditional unstructured interviews. These interviews often disadvantage neurodivergent candidates by overemphasizing social interaction and spontaneous verbal responses, which may not align with their communication styles. Structured interviews, where all candidates are asked identical questions, provide a more equitable platform by focusing on relevant skills and measurable experiences (Hendrickx, 2010). Furthermore, task-based evaluations can serve as an objective mechanism to measure a candidate's competency. For instance, presenting coding challenges, problem-solving exercises, or design projects allows neurodivergent individuals to showcase their abilities in practical, low-pressure contexts. These methods also minimize subjective interpretations by hiring managers, fostering consistency across evaluations (Table 3.1).

Another critical component of inclusive recruitment is the redesign of job descriptions. Ambiguity in job postings—such as the use of jargon or vague language—can deter neurodivergent applicants, who may struggle to decipher implied expectations. By employing clear, concise, and specific language, organizations can significantly enhance accessibility. For example, instead of requesting a *self-starter with excellent interpersonal skills,* a job posting might specify *Candidates should be capable of managing project timelines independently and collaborating with team members* via *digital tools*. Such clarity not only attracts a broader talent pool but also sets the stage for more accurate performance evaluations. Moreover, explicitly stating the availability of workplace accommodations and flexible arrangements can help reduce anxieties about the recruitment process and foster trust among potential applicants.

Inclusive recruitment efforts also benefit from partnerships with organizations specializing in neurodiversity. Collaborating with advocacy groups, such as *Specialisterne* (https://specialisterne.com/) or local neurodiversity networks, can

Table 3.1 Traditional vs. inclusive hiring practices

Aspect of recruitment	Traditional practice	Inclusive alternative
Interview format	Unstructured, conversational	Structured, skills-focused
Assessment	Verbal reasoning and social interaction	Task-based evaluations
Job description	Ambiguous, jargon-heavy	Clear, concise, and specific
Candidate pooling	General outreach	Partnerships with neurodiversity organizations

help companies tap into a well-qualified, yet underrepresented, talent pool. These partnerships often include tailored guidance on making hiring processes more accessible and connecting with neurodivergent candidates who may otherwise hesitate to apply. Additionally, neurodiversity consultants can train hiring managers and HR professionals to recognize implicit biases and adjust evaluation criteria accordingly, further embedding inclusivity into organizational practices.

Leading organizations have already begun implementing these strategies with remarkable success. Microsoft's Neurodiversity Hiring Program, for instance, replaces conventional interviews with multi-day assessment processes designed to evaluate skills in a supportive environment. This innovative approach allows candidates to interact with potential colleagues and demonstrate their abilities through practical tasks, reducing performance anxieties. Similarly, SAP's Autism at Work initiative prioritizes task-based assessments and provides candidates with detailed explanations of the recruitment process to reduce uncertainty (Austin & Pisano, 2017). Both programs illustrate how a focus on inclusive practices can yield significant benefits, including heightened team creativity, improved employee retention, and stronger overall performance.

Through these measures, organizations can build equitable hiring processes that not only diversify their workforce but also enhance their innovative capacity. A reimagined recruitment pipeline—characterized by structured assessments, transparent communication, and active outreach to neurodivergent communities—can transform hiring outcomes. By embracing the principles of inclusivity, companies are better positioned to attract, retain, and nurture neurodivergent talent—an essential step in meeting the challenges of an increasingly dynamic global economy. Furthermore, these initiatives send a clear message to stakeholders and employees alike: diversity and inclusion are not just ethical imperatives but strategic advantages in a competitive marketplace.

3.4 Effective Onboarding Practices

The onboarding process represents a pivotal opportunity to boost both individual and organizational productivity. When neurodivergent employees are effectively onboarded, they experience a smoother transition into their roles, which accelerates their ability to contribute meaningfully to team and organizational goals. Tailored onboarding practices that align with the strengths of neurodivergent individuals can result in significant productivity gains. For instance, neurodivergent employees who are given clear, structured workflows and task-specific training often adapt more quickly, reducing ramp-up time and achieving high levels of output earlier than traditional onboarding models allow (Hennekam & Follmer, 2024; Austin & Pisano, 2017). Moreover, by addressing common barriers during onboarding—such as unclear expectations or insufficient accommodations—organizations can mitigate inefficiencies and unlock the potential for higher productivity across teams (Fig. 3.2).

A personalized onboarding plan tailored to the strengths, challenges, and preferences of neurodivergent employees is a foundational element of inclusivity. Employees who experience personalized support during the onboarding phase may exhibit higher levels of engagement, confidence, and long-term retention. For instance, providing neurodivergent employees with a clear, structured overview of job expectations, organizational hierarchies, and performance metrics can reduce anxiety and enhance clarity. Leaders should prioritize transparent and direct communication, as ambiguous instructions or vague expectations may hinder the confidence of neurodivergent individuals, particularly those on the autism spectrum (Szulc, 2024). Tailoring these plans further requires collaboration with the employees themselves, encouraging open dialogue to ensure that their voices are central to shaping the process.

Equipping managers and colleagues with neurodiversity awareness training further strengthens the onboarding process. Training initiatives should focus on fostering empathy, understanding cognitive differences, and cultivating environments of psychological safety. For example, equipping teams with practical strategies to communicate with neurodivergent peers—such as utilizing concise, precise language or avoiding excessive reliance on nonverbal cues—can significantly improve collaboration and cohesion. Empirical evidence suggests that neurodiversity

Fig. 3.2 Relation between onboarding and performance

training enhances team dynamics and mitigates potential misunderstandings or biases that may arise in diverse work environments (Silberman, 2015). Moreover, such training contributes to a broader organizational culture that values and respects cognitive diversity, ensuring that neurodivergent employees feel included from the outset.

Mentorship programs are another highly effective tool in facilitating the successful onboarding of neurodivergent employees. Assigning mentors who can offer guidance, answer questions, and provide emotional support during the transition period creates a sense of security and belonging. Mentors familiar with neurodivergence can also act as advocates, assisting in navigating workplace accommodations or advocating for adjustments to workflows that align with the employee's unique strengths. Additionally, mentorship fosters knowledge transfer and builds rapport, reinforcing the neurodivergent employee's connection to the organization. When paired with peer coaching initiatives, mentorship programs can provide a multilayered support system that enhances confidence and resilience among neurodivergent hires.

Technology plays a pivotal role in enabling neurodivergent employees to thrive during onboarding. Assistive technologies such as noise-canceling headphones, task management software with visual aids, and speech-to-text tools can mitigate sensory and cognitive challenges. Quiet zones and flexible workstations further provide environments conducive to focus and productivity. Organizations that integrate such tools into their onboarding frameworks report heightened employee satisfaction and engagement, particularly among neurodivergent hires (Hendrickx, 2010). Additionally, digital tools designed to track progress and provide visualized milestones can further enhance clarity and motivation for neurodivergent employees, supporting them as they acclimate to their roles (Fig. 3.3).

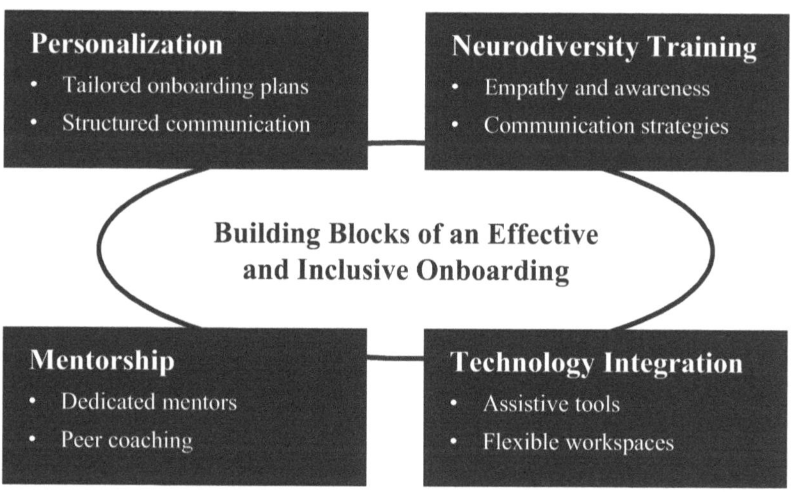

Fig. 3.3 Onboarding building blocks

Finally, the onboarding process must be iterative, involving regular check-ins to assess progress and address challenges. Leaders should schedule structured feedback sessions to solicit insights from neurodivergent employees on what aspects of the onboarding process are effective or require improvement. Tailored feedback not only allows the organization to refine its practices but also demonstrates a commitment to employee well-being and development. By adopting an iterative approach, organizations can continuously enhance their onboarding strategies, ensuring that they remain responsive to the evolving needs of neurodivergent employees. In parallel, managers should create open channels for informal feedback, empowering neurodivergent employees to share their experiences without the pressure of formal evaluations. Over time, this iterative process fosters a dynamic onboarding framework that evolves with organizational goals and employee requirements, creating a robust foundation for long-term success.

3.5 Measuring Success

Measuring the success of onboarding programs tailored to neurodivergent employees is critical to ensuring the long-term viability and effectiveness of these initiatives. Success metrics not only provide a roadmap for continuous improvement but also offer tangible evidence of the value that neurodivergent employees contribute to organizational objectives. An effective measurement strategy should encompass both quantitative and qualitative approaches, capturing the multifaceted impact of neurodiversity in the workplace.

A primary quantitative metric for measuring onboarding success is employee retention and/or (early) attrition. Retention rates among neurodivergent hires can provide insights into whether the onboarding process adequately meets their needs. Research indicates that inclusive onboarding practices can significantly improve retention rates, as employees who feel supported are more likely to remain with the organization (Hennekam & Follmer, 2024). Organizations can compare retention rates of neurodivergent employees with industry benchmarks or with neurotypical peers to identify gaps and opportunities for improvement. Moreover, tracking retention over multiple years can offer insights into the sustainability of onboarding practices, providing data on long-term integration and employee satisfaction.

Another key metric is job performance, assessed through regular performance evaluations and achievement of predefined goals. For example, neurodivergent employees in roles requiring high attention to detail, such as software testing or data analysis, may demonstrate superior outcomes when their unique strengths are effectively harnessed. Performance data can help organizations assess whether tailored onboarding practices enhance employee contributions to key performance indicators (KPIs) such as productivity, error reduction, and innovation. In addition to individual performance, team-based performance metrics can reveal how neurodivergent employees collaborate within cross-functional teams, further showcasing their impact.

Employee engagement is a crucial qualitative metric that reflects the emotional and cognitive connection neurodivergent employees have with their work and the organization. Surveys designed to assess engagement levels can include questions on job satisfaction, feelings of inclusion, and perceptions of support. A particularly valuable tool in measuring engagement is the Employee Net Promoter Score (eNPS), which gauges the likelihood of employees recommending their organization as a great place to work. The eNPS metric provides a clear, quantifiable indicator of overall employee sentiment and can be segmented to analyze the specific experiences of neurodivergent employees (Reichheld, 2003). Studies show that neurodivergent employees who receive targeted onboarding and accommodations report higher levels of engagement and workplace satisfaction, which correlates with enhanced team cohesion and organizational loyalty (Silberman, 2015). To gain deeper insights, engagement metrics should be segmented by departments or teams to identify where interventions have been most effective.

Another valuable qualitative metric is feedback from neurodivergent employees regarding their onboarding experience. Regular feedback sessions, both formal and informal, can provide actionable insights into the efficacy of onboarding practices. This feedback helps identify specific challenges employees face and highlights areas for improvement. Open-ended questions in surveys or one-on-one meetings can encourage employees to share nuanced perspectives that might not emerge through standardized metrics alone. Additionally, longitudinal studies that track feedback over time can reveal how onboarding practices evolve to better meet employee needs.

Organizations should also measure the broader cultural impact of their neurodiversity initiatives. Metrics such as team dynamics, collaboration quality, and innovation levels can indicate how well neurodivergent employees are integrated into the workplace. For instance, tracking the frequency and outcomes of cross-functional projects involving neurodivergent team members can provide insights into the value of cognitive diversity in driving innovative solutions (Hendrickx, 2010). Furthermore, assessing shifts in employee attitudes toward neurodiversity through periodic culture surveys can gauge the overall acceptance and inclusion of neurodivergent individuals within the organization (Fig. 3.4).

To visualize these success metrics, organizations can employ dashboards that integrate retention rates, performance outcomes, and engagement scores, allowing HR teams and leadership to monitor progress in real time. Dashboards can also be segmented by timeframes, such as quarterly or yearly views, to reveal trends. Additionally, a table comparing pre- and post-implementation data on these metrics can demonstrate the tangible benefits of neurodivergent onboarding programs. By providing both high-level summaries and granular details, these tools enable organizations to align their neurodiversity strategies with broader business goals.

3.6 Conclusion

The integration of neurodivergent employees into the workforce is not merely a matter of equity but also a strategic imperative for organizations seeking to drive innovation and performance. As evidenced by initiatives like Microsoft's Neurodiversity Hiring Program, tailored onboarding practices can unlock the full

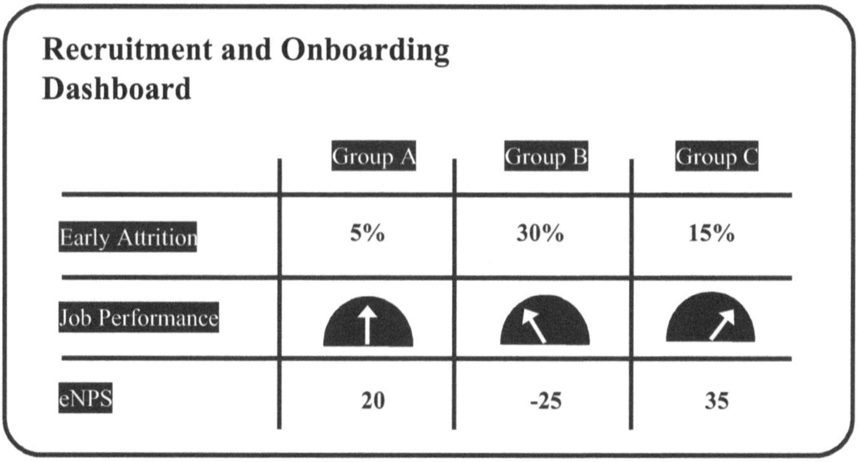

Fig. 3.4 Example of a (simplified) dashboard

potential of neurodivergent talent, resulting in measurable gains in productivity, team dynamics, and retention.

Effective onboarding requires a multifaceted approach, encompassing personalized support, structured workflows, and the integration of assistive technologies. By addressing the specific needs of neurodivergent employees, organizations can foster environments that enable them to thrive. Metrics such as employee retention, engagement, and performance provide valuable insights into the success of these practices and serve as a foundation for continuous improvement.

Beyond individual outcomes, neurodiversity initiatives contribute to broader organizational goals by promoting cognitive diversity and inclusive cultures. By embracing the strengths of neurodivergent employees, organizations position themselves as leaders in innovation and equity. The lessons drawn from this chapter underscore the importance of systemic change in recruitment and onboarding processes to ensure that all employees, regardless of cognitive differences, can contribute meaningfully to organizational success.

The journey toward neurodiversity inclusion is ongoing, requiring sustained commitment and adaptation. Organizations that invest in these efforts will not only enhance their competitive advantage but also create workplaces where all employees feel valued and empowered to excel.

3.7 Case Study: SAP's Autism at Work Program

SAP's Autism at Work program provides a compelling example of how organizations can harness the potential of neurodivergent employees through targeted recruitment and onboarding practices. Launched in 2013, the initiative was designed to address both the underrepresentation of individuals with autism in the workforce and the need for specialized skills in areas such as software testing, quality assurance, and cybersecurity (Austin & Pisano, 2017, Hennekam & Follmer, 2024).

The program begins with a recruitment process tailored to the needs of neurodivergent candidates. Instead of relying on traditional interviews, which often emphasize verbal communication and social interaction, SAP employs a more inclusive approach. Candidates participate in workshops that assess their technical and problem-solving abilities in collaborative, low-pressure environments. This method not only allows candidates to demonstrate their strengths but also reduces the anxiety often associated with conventional hiring practices (SAP, 2025).

Once hired, SAP's onboarding process focuses on creating an environment where neurodivergent employees can thrive. The company provides comprehensive support, including job coaches who help new hires navigate workplace expectations and adapt to their roles. Additionally, SAP offers tailored accommodations, such as flexible work schedules, sensory-friendly workspaces, and assistive technologies, to ensure that employees can work comfortably and effectively.

The impact of the Autism at Work program has been significant. Neurodivergent employees at SAP have excelled in roles that require high levels of precision and pattern recognition, contributing to innovations in software development and data analysis. For example, teams that include neurodivergent members have reported improved problem-solving capabilities and enhanced quality assurance outcomes. Moreover, the program has fostered a more inclusive organizational culture, increasing awareness and understanding of neurodiversity across the company (Austin & Pisano, 2017).

SAP's success with Autism at Work underscores the value of investing in neurodiversity initiatives. By rethinking traditional hiring and onboarding practices, the company has not only tapped into a previously underutilized talent pool but also enhanced its overall performance and innovation capacity.

References

Austin, R. D., & Pisano, G. P. (2017). Neurodiversity as a competitive advantage. *Harvard Business Review*.

Hendrickx, S. (2010). *The adolescent and adult neuro-diversity handbook: Asperger syndrome, ADHD, dyslexia, Dyspraxia and Related Conditions*. Jessica Kingsley Publishers.

Hennekam, S., & Follmer, K. (2024). Neurodiversity and HRM: a practice-based review and research agenda. *Equality, Diversity and Inclusion, 43*(7), 1119–1129.

Hewlett, S. A., Marshall, M., & Sherbin, L. (2013). How diversity drives innovation. *Harvard Business Review*.

Reichheld, F. F. (2003). The one number you need to grow. *Harvard Business Review, 81*(12), 46–54.

SAP. (2025). Autism at work program. Retrieved January 25, 2025. https://www.sap.com/discover/autism-inclusion/index.html.

Silberman, S. (2015). *NeuroTribes: The legacy of autism and the future of neurodiversity*. Avery.

Singer, J. (2016). *Neurodiversity: The birth of an idea*. Kindle.

Szulc, J. M. (2024). Leading with understanding: cultivating positive relationships between neurotypical leaders and neurodivergent employees. *Employee Relations: The International Journal, 46*(9), 97–114.

4 Innovative Job and Workplace Design: Hybrid Models and Flexibility for Diverse Teams

4.1 Introduction: The Evolving Needs of Workplace Design

The design of jobs and workplaces has undergone a significant transformation in recent years, driven by the evolving demands of a diverse workforce, technological advancements, and societal shifts. Central to this transformation is the growing recognition of the need for inclusive and flexible work environments that cater to a wide range of employee needs, including those of neurodivergent individuals. As organizations adapt to rapid changes in the nature of work, hybrid models and flexible approaches have emerged as pivotal strategies to foster engagement, productivity, and inclusivity.

Traditionally, workplace design focused on standardization and uniformity, often emphasizing a one-size-fits-all approach to physical spaces and job roles. This conventional framework, however, often overlooked the unique needs of employees, particularly those with neurodivergent conditions such as autism spectrum disorder (ASD), ADHD, or dyslexia. For neurodivergent employees, rigid schedules, sensory-unfriendly environments, and inflexible job structures can create significant barriers to performance and well-being (Szulc, 2024). Recent shifts in workplace practices, including the rise of hybrid work and flexible schedules, have opened new opportunities to address these challenges.

Hybrid work models, which combine in-office and remote work, exemplify the growing emphasis on flexibility in workplace design. Research has shown that hybrid models enhance employee satisfaction and productivity while reducing stress and burnout by enabling employees to customize their work environments (BCG, 2020). For neurodivergent employees, the ability to work in environments that accommodate their sensory and cognitive needs—whether at home or in the office—significantly enhances their performance. However, hybrid work also introduces challenges, including the need to maintain cohesion, collaboration, and

M. Helmold, M. Martensen, *Leadership in Neurodiversity and Neurodivergence*, Diversity and Inclusion Research, https://doi.org/10.1007/978-3-032-09343-1_4

inclusion across dispersed teams. Leaders must navigate these complexities to ensure that hybrid models benefit all employees, regardless of their neurocognitive profiles.

Flexibility extends beyond where work takes place to include how and when it is performed. Flexible working hours allow employees to align their work schedules with their peak productivity times and personal needs. For instance, individuals with ADHD often experience productivity spikes during unconventional hours, such as late evenings, making traditional nine-to-five schedules less effective. Similarly, flexible job structures that emphasize outcomes over rigid processes can empower neurodivergent employees to leverage their unique strengths, such as hyper-focus or creative problem-solving.

The integration of flexibility into workplace design is not merely a response to employee preferences; it is a strategic imperative for organizations aiming to remain competitive in a dynamic and uncertain environment. Flexibility enables organizations to attract and retain diverse talent, fostering innovation and resilience in the face of complex challenges. Furthermore, inclusive workplace design aligns with broader trends toward equity and diversity, reinforcing organizational commitments to social responsibility and ethical leadership (Dennison, 2024; Rollnik-Sadowska & Grabinska, 2024). In addition to flexibility and hybrid models, technology plays a critical role in modern workplace design. Collaboration tools, such as Slack or Microsoft Teams, have become indispensable in enabling seamless communication across hybrid teams. For neurodivergent employees, assistive technologies, such as noise-canceling headphones, text-to-speech software, or task management apps, may further enhance their ability to thrive in both remote and in-office settings. By integrating these technologies into workplace strategies, organizations can create environments where employees feel supported and empowered to excel.

The evolving needs of workplace design thus reflect a broader shift toward prioritizing employee well-being, inclusion, and flexibility. Hybrid work models and flexible structures are no longer optional features but essential components of forward-thinking organizations. For neurodivergent employees, these innovations provide opportunities to overcome traditional barriers and contribute meaningfully to organizational success. As subsequent sections will explore, effectively implementing these approaches requires a nuanced understanding of employee needs, strategic leadership, and a commitment to continuous improvement.

4.2 Foundational Elements of Inclusive Job Design

The concept of job design lies at the heart of creating workplaces that empower employees to perform at their best while fostering inclusion and well-being. Effective job design not only aligns roles with organizational goals but also tailors tasks and environments to the unique needs of employees. For neurodivergent individuals, this alignment is especially critical, as it ensures that job roles leverage their strengths while minimizing barriers that may hinder performance.

At its core, inclusive job design incorporates principles such as role clarity, autonomy, and flexibility. Role clarity refers to the clear definition of tasks, responsibilities, and expectations. Research indicates that ambiguity in job roles can exacerbate anxiety and stress, particularly for neurodivergent employees who thrive on predictability and structure (Hendrickx, 2010). By providing detailed role descriptions and clear communication, employers can create an environment where all employees feel confident in their responsibilities.

Autonomy, another cornerstone of effective job design, empowers employees to make decisions about how they approach their work. For neurodivergent employees, autonomy can enable the use of personalized strategies that align with their unique cognitive styles. For instance, individuals with dyslexia may prefer using assistive technologies for writing tasks, while those with ADHD may excel when allowed to break tasks into smaller, manageable steps (Austin & Pisano, 2017). Providing autonomy fosters a sense of ownership and engagement, which, in turn, enhances productivity and job satisfaction.

Flexibility, too, is increasingly recognized as a critical component of inclusive job design, especially in the context of hybrid and remote work models. Flexible work arrangements allow employees to adjust their schedules or work environments to better suit their needs. For example, employees with sensory sensitivities may benefit from the option to work remotely, where they can control lighting and noise levels. Similarly, flexible hours can accommodate those whose peak productivity occurs outside traditional office hours (Hennekam & Follmer, 2024). Flexibility not only supports neurodivergent employees but also benefits the broader workforce by fostering a culture of trust and adaptability.

A notable framework for understanding job design is Hackman and Oldham's Job Characteristics Model (JCM), which identifies five core job dimensions: skill variety, task identity, task significance, autonomy, and feedback (Hackman & Oldham, 1976). Inclusive job design leverages these dimensions to create roles that are meaningful and engaging for all employees. For neurodivergent employees, tailoring these dimensions can be particularly impactful. For instance, providing regular and constructive feedback helps individuals with autism spectrum disorder (ASD) understand how their contributions align with organizational goals, while roles that emphasize task significance can boost motivation and self-esteem. Table 4.1 illustrates how the principles of role clarity, autonomy, and flexibility

Table 4.1 Alignment of inclusive job design principles with JCM dimensions

Inclusive principle	JCM dimension	Application in job design
Role clarity	Task identity	Clear articulation of responsibilities and expected outcomes
Autonomy	Autonomy	Allowing employees to customize workflows and decision-making
Flexibility	Skill variety	Offering diverse tasks and adjustable schedules to suit employee needs
Feedback	Feedback	Providing timely, specific, and constructive feedback to employees

align with the core dimensions of the Job Characteristics Model to support inclusive job design.

Inclusive job design also requires an understanding of how physical and digital workspaces intersect with job roles. For example, physical environments should accommodate diverse sensory needs, while digital tools must support collaboration and productivity. Organizations such as SAP and Microsoft have demonstrated the value of integrating inclusive practices into job design. Microsoft's Neurodiversity Hiring Program, for instance, emphasizes skill-based assessments and tailored job roles to align with the strengths of neurodivergent candidates (Microsoft, 2025). By doing so, these organizations not only support employee well-being but also harness unique talents to drive innovation.

Hence, foundational elements such as role clarity, autonomy, and flexibility form the bedrock of inclusive job design. By incorporating these principles, organizations can create work environments that not only accommodate neurodivergent employees but also foster engagement and productivity across the workforce. As the following sections will explore, integrating hybrid models and flexible organizational practices further enhances the ability of organizations to meet diverse employee needs while achieving strategic objectives.

4.3 Hybrid Work: A New Standard

The hybrid work model has gained prominence as a revolutionary approach to organizational design, balancing on-site and remote work to meet the demands of a modern and diverse workforce. While its benefits include increased flexibility, employee satisfaction, and productivity, hybrid work also introduces complexities in maintaining team cohesion and fostering inclusion. For neurodivergent employees, hybrid work can create tailored opportunities for optimal performance, provided that organizations implement supportive structures and address inherent challenges.

4.3.1 Models of Hybrid Work

Hybrid work does not represent a single standard but rather encompasses a broad spectrum of models that cater to the diverse needs of organizations and individuals. These models include fully collocated arrangements, alternating schedules, and more flexible options like on-demand attendance or work-from-anywhere setups. Fully collocated models, where employees are required to work entirely on-site, offer the advantage of spontaneous collaboration and immediate access to resources but significantly limit flexibility. This model is particularly suited for roles that demand frequent interpersonal interactions or reliance on on-site equipment (Fig. 4.1).

Alternating on-site schedules provide a structured balance by allowing employees to rotate between in-office and remote work. This approach fosters team

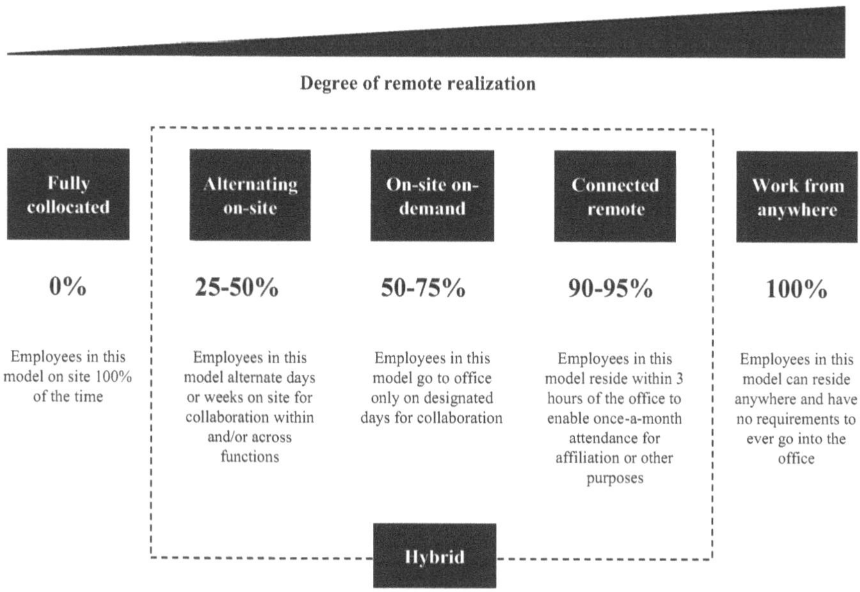

Fig. 4.1 Spectrum of remote work (BCG, 2020)

collaboration while offering some degree of personal flexibility, making it an effective choice for roles that blend teamwork with independent tasks. On-demand models, in which employees attend the office only for specific tasks, meetings, or collaborative efforts, align closely with the principle of purposeful work attendance. These models minimize unnecessary commutes while maintaining opportunities for meaningful in-person engagement.

Another innovative hybrid approach is the connected remote model, which allows employees to reside near the workplace and attend the office on rare occasions when required. This model balances the benefits of proximity to the workplace with the flexibility of predominantly remote work. Finally, work-from-anywhere arrangements grant employees the autonomy to choose their location of work without any fixed requirement for in-office presence. This level of freedom is particularly appealing to individuals who prioritize independence but may reduce opportunities for informal interactions and team-building activities.

These hybrid work models reflect the diverse requirements of both organizations and employees, each offering unique trade-offs in terms of collaboration, flexibility, and inclusion. For example, fully collocated models emphasize teamwork and cohesion at the expense of adaptability, while work-from-anywhere setups provide maximum flexibility but require innovative strategies to maintain organizational culture. Table 4.2 illustrates the key characteristics of these hybrid work models, emphasizing their benefits and challenges across various dimensions.

Table 4.2 Comparison of hybrid work models

Hybrid model	Key characteristics	Benefits	Challenges
Fully collocated	Entirely on-site; emphasis on physical presence and spontaneous collaboration	High team cohesion and access to on-site resources	Limited flexibility; may not accommodate neurodivergent needs
Alternating on-site	Rotation between in-office and remote schedules	Balance between flexibility and collaboration; ideal for team-focused tasks	Coordination of schedules can be complex
On-site on-demand	Office attendance only for specific tasks or meetings	Reduces unnecessary commutes; supports focused, purposeful interactions	Limited opportunities for spontaneous collaboration
Connected remote	Predominantly remote with occasional in-office presence	Supports flexibility while maintaining proximity to the workplace	May risk employee isolation without regular team engagement
Work-from-anywhere	Full flexibility to work from any location, with no requirement to attend office	Maximum autonomy; suited to roles emphasizing individual output	Reduced in-person engagement may hinder team-building and inclusion efforts

4.3.2 Hybrid Work and Neurodiverse Teams

Hybrid work offers unique advantages for neurodivergent employees by enabling them to customize their work environments and schedules to align with their individual needs. For instance, employees with autism spectrum often experience sensory sensitivities that make traditional office environments overwhelming. The ability to work remotely allows them to create controlled settings with minimal sensory distractions, improving focus and overall productivity. Similarly, individuals with ADHD may benefit from hybrid work setups that enable them to structure their day around their peak productivity times, often outside traditional office hours (Hennekam & Follmer, 2024). Furthermore, hybrid work models can empower neurodivergent employees to integrate wellness strategies into their daily routines, such as engaging in physical activity or mindfulness practices during breaks, further enhancing their overall well-being and productivity.

Despite these advantages, hybrid work also introduces challenges for neurodivergent employees. Working remotely can exacerbate feelings of isolation and disconnection from the team, potentially reducing engagement and inclusion. This issue is particularly pronounced in scenarios where communication practices are inconsistent or fail to account for neurodivergent needs. A lack of tailored support for remote employees may further hinder collaboration, creating barriers to performance. To address these issues, organizations must implement strategies that foster connection and ensure equity across hybrid teams. Regular check-ins with managers, virtual team-building activities, and the integration of assistive technologies can play a critical role in supporting neurodivergent employees in hybrid setups. These efforts should be complemented by providing training for managers to effectively engage neurodivergent employees, emphasizing empathy, adaptability, and clarity in communication.

Organizations that successfully integrate neurodivergent employees into hybrid work environments often invest in customized workflows, enabling employees to perform tasks at their best. For instance, companies may introduce structured flexibility, allowing employees to choose not only where they work but also how their tasks are distributed over the day. Microsoft's Neurodiversity Hiring Program provides a valuable example, emphasizing flexibility in job roles and work arrangements, empowering neurodivergent employees to thrive in both on-site and remote settings. By pairing these practices with inclusive team-building initiatives, such as neurodiversity awareness training and collaborative workshops, organizations can create an environment where hybrid work becomes a tool for innovation and inclusivity. These practices demonstrate the potential of hybrid work not only to enhance inclusivity but also to drive organizational success by unlocking the full potential of diverse talent.

4.3.3 Challenges and Solutions for Hybrid Work

While hybrid work offers numerous benefits, it also poses challenges that must be addressed to ensure its success. One major challenge is maintaining team cohesion in a dispersed work environment. Employees who work primarily remotely may feel disconnected from their on-site colleagues, leading to reduced collaboration and a sense of exclusion. To overcome this, organizations can implement regular team-wide updates, virtual meetings, and periodic in-person gatherings to bring all members together, regardless of their location. These interactions foster alignment and build trust across hybrid teams, ensuring that all employees feel valued and included in collective decision-making processes. Additionally, encouraging informal virtual interactions, such as casual coffee chats or team-building exercises, can help maintain camaraderie in hybrid settings.

Another challenge lies in ensuring equitable access to resources. Remote employees may lack the tools and equipment readily available to their on-site counterparts, creating disparities in their ability to perform tasks effectively. Organizations should prioritize providing all employees with necessary resources, including ergonomic workstations, reliable internet connections, and advanced digital collaboration tools. By offering stipends for home office setups or creating shared hubs for occasional use, companies can bridge these gaps. Additionally, investing in comprehensive training programs for both on-site and remote employees ensures consistent skill development and fosters a level playing field where everyone has the tools and knowledge to succeed.

A further challenge involves addressing collaboration gaps that arise in hybrid work environments. Spontaneous interactions, often a hallmark of in-office settings, are difficult to replicate remotely. These unplanned discussions frequently spark innovation and build team rapport. To counter this, organizations can design physical spaces that encourage meaningful collaboration during on-site days, such as open meeting areas, breakout rooms for brainstorming, and quiet zones for focused work. Moreover, leveraging digital platforms like Slack, Microsoft Teams, and Asana can facilitate asynchronous collaboration, enabling team members to contribute ideas and feedback at their convenience. Integrating hybrid-friendly workflows—such as using virtual whiteboards or task-tracking systems—can further streamline the collaborative process and ensure that every team member, regardless of location, remains engaged and productive (BCG, 2020).

By addressing these challenges with intentional strategies, hybrid work can evolve into a model that not only enhances flexibility but also strengthens team cohesion and equitable collaboration. Organizations that proactively invest in resources, foster communication, and adapt physical and digital environments will be better positioned to maximize the benefits of hybrid work while creating inclusive, innovative workplaces for diverse teams. For neurodivergent employees, hybrid models provide opportunities to overcome traditional barriers and thrive in personalized environments. However, realizing the full potential of hybrid work requires organizations to address its challenges through intentional strategies, robust communication, and equitable resource distribution. By doing so, they can create hybrid work environments that empower all employees to succeed, regardless of their location or neurocognitive profile.

4.4 Flexible Working Hours and Flexible Organization

As discussed, the modern workplace is undergoing a profound transformation, driven by shifting employee expectations, technological advancements, and the increasing need for adaptability. Among the most significant developments is the rise of flexible working hours and organizational flexibility. These approaches provide employees with greater autonomy over when and how they work, fostering a sense of empowerment and enhancing productivity. For neurodivergent employees, these flexible arrangements offer opportunities to optimize their schedules and work environments to align with their unique strengths, while organizations benefit from improved engagement, innovation, and retention.

Flexible working hours, a cornerstone of modern organizational design, allow employees to adapt their work schedules to better fit their personal needs and productivity patterns. Traditional nine-to-five schedules, i.e., rigid working hours (see Fig. 4.2), often fail to accommodate the diverse rhythms of human productivity, particularly for individuals with neurodivergent conditions such as ADHD or autism spectrum disorder. Many neurodivergent individuals perform best during unconventional hours, such as early mornings or late evenings, when they can work with fewer distractions. By offering flexibility in scheduling, organizations empower employees to work at their peak, resulting in higher quality outputs and reduced stress.

Moreover, flexible working hours can significantly improve work–life balance, a factor closely linked to employee satisfaction and well-being. Employees with

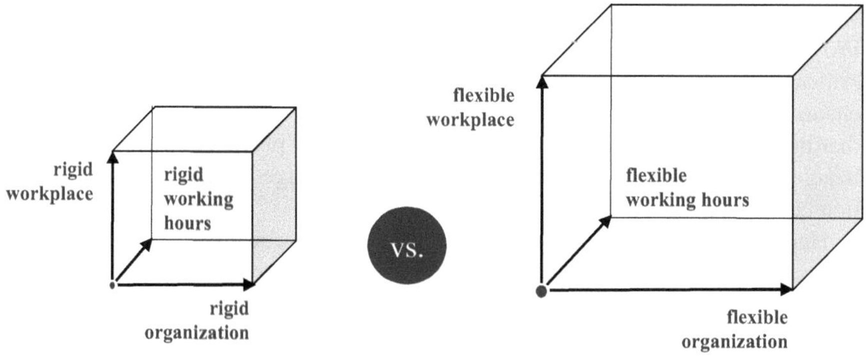

Fig. 4.2 Rigid vs. flexible work policies (Trost, 2020)

caregiving responsibilities or other personal commitments benefit from the ability to tailor their work schedules around these obligations. For example, an employee managing both work and caregiving for a family member may find that a flexible schedule reduces the risk of burnout and enhances their overall productivity. These arrangements also support neurodivergent employees who may require breaks to manage sensory overload or other challenges, enabling them to maintain focus and energy throughout the day.

Organizational flexibility extends beyond individual schedules to encompass broader structural adaptations. Flexible organizations prioritize agility, fostering cultures that emphasize collaboration, adaptability, and outcome-driven performance. By moving away from rigid hierarchies and fixed roles, these organizations create environments where employees can thrive, regardless of their cognitive profiles. For neurodivergent employees, flexible roles may involve adjustments to responsibilities or workflows that align with their specific strengths. For instance, an employee with exceptional pattern-recognition skills may excel in data analysis, while another with strong creative abilities may thrive in design-oriented tasks. These adjustments not only improve individual performance but also contribute to organizational success by leveraging the full potential of diverse talents.

Flexible working hours and organizational structures also require robust mechanisms for evaluation and continuous improvement. Measuring the success of these practices involves assessing their impact on employee satisfaction, productivity, and overall organizational performance. Outcome-based key performance indicators (KPIs) provide a useful framework for evaluating the effectiveness of flexible work arrangements. For instance, employee engagement surveys can capture feedback on work–life balance and job satisfaction, while productivity metrics can track the quality and timeliness of deliverables. Additionally, retention rates and talent acquisition metrics can offer insights into the broader impact of flexibility on organizational culture and competitiveness (Dennison, 2024).

To implement and sustain flexible working practices, leadership must play a pivotal role in setting the tone and aligning organizational policies with employee needs. Leaders should prioritize transparency and trust, creating an environment where employees feel confident in their ability to navigate flexible arrangements. Providing training for managers on how to lead flexible teams effectively is equally critical. For example, managers must learn to balance accountability with empathy, ensuring that employees remain productive while feeling supported. Organizations that invest in these leadership competencies are better positioned to maximize the benefits of flexibility while addressing potential challenges such as miscommunication or reduced team cohesion.

Hence, flexible working hours and organizational structures represent a fundamental shift in how work is designed and managed. By embracing these practices, organizations can create environments that support the diverse needs of their workforce, including neurodivergent employees. Flexibility fosters not only individual well-being and productivity but also organizational resilience and innovation. As workplaces continue to evolve, the integration of flexible practices will remain a cornerstone of successful, inclusive organizations.

4.5 Designing Sensory-Friendly and Adaptive Work Environments

Work environments play a critical role in shaping employee well-being, productivity, and engagement. For neurodivergent employees, the design of the workplace can significantly impact their ability to thrive. Sensory-friendly and adaptive work environments are essential for accommodating the diverse sensory and cognitive needs of employees, enabling them to perform at their best. As organizations strive to create inclusive workplaces, they must consider physical, technological, and organizational adaptations that promote accessibility and comfort for all.

A sensory-friendly workplace takes into account the diverse sensory sensitivities that many employees experience, particularly those with conditions such as autism spectrum disorder (ASD), ADHD, or sensory processing disorder (SPD). Common environmental factors such as harsh lighting, loud noises, or cluttered spaces can be overwhelming for these individuals, leading to stress and decreased productivity. To address these challenges, organizations can implement practical modifications such as adjustable lighting, soundproofing, and ergonomic furniture. For instance, providing employees with dimmable lights or access to noise-canceling headphones allows them to customize their environment to suit their sensory needs.

Adaptive work environments go beyond sensory considerations to include flexible layouts and dynamic spaces that can be tailored to individual and team requirements. Open-plan offices, while designed to promote collaboration, often create distractions and discomfort for many employees, including neurodivergent individuals. Incorporating quiet zones, private booths, and flexible seating arrangements enables employees to choose spaces that align with their preferred working styles. Such adaptive spaces may not only improve focus and productivity but also foster a sense of autonomy and control among employees.

Technology, too, plays a vital role in enhancing sensory-friendly and adaptive work environments. Assistive technologies such as text-to-speech software, screen

Table 4.3 Features of sensory-friendly and adaptive work environments

Feature	Description	Benefits
Adjustable lighting	Dimmable or natural lighting options	Reduces sensory discomfort and eye strain
Noise control	Soundproofing, noise-canceling headphones	Minimizes distractions and supports focus
Flexible layouts	Quiet zones, private booths, collaborative areas	Accommodates diverse working styles and needs
Assistive technologies	Tools such as text-to-speech software and screen magnifiers	Enhances accessibility for employees with sensory or cognitive challenges
Virtual reality design tools	Simulation of sensory-friendly workspaces	Enables testing and refinement of workplace modifications

magnifiers, and wearable devices can support employees with visual or auditory impairments, as well as those with neurodivergent conditions. Collaboration tools like Slack and Microsoft Teams provide alternative communication channels, enabling employees to participate in discussions in ways that feel most comfortable to them. Additionally, virtual reality (VR) tools are increasingly being used to design and simulate sensory-friendly spaces, allowing organizations to test and refine their workplace modifications before implementation. Table 4.3 outlines key features of sensory-friendly and adaptive work environments, highlighting their benefits for employee well-being and organizational outcomes.

Organizations that prioritize sensory-friendly and adaptive environments also foster a culture of inclusion and psychological safety. When employees feel that their needs are understood and respected, they are more likely to engage fully and contribute to organizational success. For example, SAP's Autism at Work program incorporates sensory-friendly design elements such as quiet zones and individualized workspaces, enabling neurodivergent employees to excel. Feedback from employees in such programs highlights increased satisfaction, productivity, and a sense of belonging. Thus, designing sensory-friendly and adaptive work environments is a critical step toward creating inclusive workplaces that empower all employees. By addressing sensory sensitivities, providing flexible spaces, and leveraging technology, organizations can enhance well-being and productivity while fostering a culture of inclusion. As the workforce becomes increasingly diverse, these practices will remain central to organizational success.

4.6 Conclusion and Recommendations

The evolving landscape of work requires organizations to adopt innovative and inclusive strategies to meet the diverse needs of their workforce. As demonstrated throughout this chapter, embracing hybrid models, flexible work arrangements, sensory-friendly environments, and adaptive organizational practices is not merely a response to employee demands but a strategic imperative for fostering productivity, engagement, and innovation. These approaches, when effectively implemented, empower employees, particularly those who are neurodivergent, to thrive and contribute meaningfully to organizational success.

Hybrid work models have emerged as a transformative framework, blending on-site and remote work to offer flexibility and autonomy. For neurodivergent employees, hybrid setups provide the opportunity to work in environments that cater to their sensory and cognitive preferences, reducing stress and enhancing productivity (Austin & Pisano, 2017). However, hybrid models must be carefully managed to ensure equitable access to resources and maintain team cohesion across dispersed teams. Organizations that succeed in addressing these challenges will unlock the full potential of hybrid work.

Flexible working hours and organizational practices further enhance inclusivity by allowing employees to align their work schedules and responsibilities with their peak productivity times and personal needs. These practices not only improve work–life balance but also foster a culture of trust and autonomy, which are critical for employee satisfaction and retention. Additionally, flexibility enables organizations to attract and retain diverse talent, enhancing innovation and resilience in a competitive market.

Designing sensory-friendly and adaptive work environments is another cornerstone of inclusive organizational design. Physical adaptations such as adjustable lighting, noise control, and flexible layouts, combined with assistive technologies, create spaces where employees can work comfortably and effectively. For neurodivergent employees, these modifications reduce barriers and may enable them to perform at their best, contributing to a more inclusive and innovative workplace culture. Organizations that prioritize sensory-friendly design not only improve individual well-being but also enhance overall organizational outcomes by fostering engagement and collaboration.

To implement these strategies effectively, organizations must adopt a holistic approach that integrates physical, technological, and cultural adaptations. Leadership plays a pivotal role in driving this transformation by fostering a culture of inclusion, investing in resources, and ensuring that policies align with employee needs. Training managers to support neurodiverse teams, implementing feedback mechanisms, and using metrics to evaluate success are critical steps in sustaining these practices.

4.7 Case Study

Ernst & Young (EY), one of the world's largest professional services firms, has long been recognized for its commitment to diversity and inclusion. In recent years, the organization has made significant strides in creating an inclusive work environment for neurodivergent employees. EY's Neuro-Diverse Centers of Excellence (NCoE) initiative, launched in 2016, serves as a compelling example of how organizations can successfully integrate neurodiverse talent while driving innovation and business success (Ernst & Young, 2021; Carroll, 2024).

The NCoE initiative was designed to recruit and support individuals with neurodivergent conditions, such as autism spectrum disorder (ASD), ADHD, and dyslexia. EY recognized the unique strengths of neurodivergent individuals, including exceptional attention to detail, pattern recognition, and problem-solving skills. The organization also acknowledged the barriers that these individuals often face in traditional hiring and workplace environments, including sensory sensitivities and rigid processes that fail to account for their needs. By addressing these challenges, EY created an environment where neurodivergent employees could thrive (Ernst & Young, 2025a, 2025b).

One of the key features of the NCoE initiative is its focus on job customization and tailored support. EY reimagined its hiring practices to emphasize candidates' skills and potential rather than traditional interview formats, which often disadvantage neurodivergent individuals. For instance, the company implemented skill-based assessments and collaborative exercises that allowed candidates to demonstrate their abilities in real-world scenarios. Once hired, neurodivergent employees received individualized support, including sensory-friendly workspaces, mentorship programs, and flexible work arrangements.

EY also adapted its workplace design to accommodate the diverse sensory and cognitive needs of its employees. In its NCoE offices, the organization introduced quiet zones, adjustable lighting, and ergonomic furniture to create a comfortable and productive environment. These modifications were particularly beneficial for employees with sensory sensitivities, enabling them to focus and perform at their best (Ernst & Young, 2021). Additionally, EY leveraged assistive technologies, such as screen readers and speech-to-text tools, to enhance accessibility for employees with different needs.

The impact of the NCoE initiative has been significant, both for EY and for its employees. Internally, neurodivergent employees have excelled in roles that require high levels of precision, such as data analysis, cybersecurity, and process automation. Externally, the initiative has strengthened EY's reputation as a leader in diversity and inclusion, attracting top talent and fostering client trust. According to EY, teams that include neurodivergent employees have demonstrated higher levels of innovation and efficiency, underscoring the business value of inclusion (Carroll, 2024).

EY's experience highlights the importance of leadership and collaboration in driving inclusive practices. The success of the NCoE initiative was made possible by the commitment of senior leaders, who championed diversity and inclusion as core organizational values. EY also partnered with external organizations specializing in neurodiversity, such as Autism Speaks and Specialisterne, to develop best practices and build a sustainable model for inclusion. In conclusion, EY's Neuro-Diverse Centers of Excellence initiative demonstrates how organizations can harness the strengths of neurodiverse talent to achieve both social and business objectives. By creating inclusive hiring practices, adapting work environments, and providing tailored support, EY has not only empowered its employees but also gained a competitive edge in the market.

References

Austin, R. D., & Pisano, G. P. (2017). Neurodiversity as a competitive advantage. *Harvard Business Review*. Retrieved January 20, 2025. https://hbr.org/2017/05/neurodiversity-as-a-competitive-advantage.

BCG. (2020). Remote work works: So where do we go from here? Retrieved January 20, 2025. https://www.bcg.com/publications/2020/remote-work-works-so-where-do-we-go-from-here.

Carroll, L. (2024). How EY is focusing on neurodiverse talent—and why it benefits everyone. Retrieved January 5, 2025. https://www.bbc.com/worklife/article/20240320-ey-karyn-twaronite-neurodiversity-bbc-executive-interview.

Dennison, K. (2024). The benefits of neurodiversity in leadership. *Forbes*.

Ernst & Young (2025a). Neurodiversity Centres of Excellence. Retrieved January 20, 2025. https://www.ey.com/en_ca/insights/consulting/neurodiversity-centres-of-excellence-ey-canada.

Ernst & Young (2025b). Neurodiversity-powered transformation. Retrieved January 20, 2025. https://www.ey.com/en_us/services/innovation/neurodiversity-powered-transformation.

Ernst & Young. (2021). EY launches first Neuro-Diverse Centre of Excellence in the UK to boost client innovation. Retrieved 20.01.2025. https://www.ey.com/en_uk/newsroom/2021/07/ey-launches-first-neuro-diverse-centre-of-excellence-in-the-uk-to-boost-client-innovation.

Hackman, J. R., & Oldham, G. R. (1976). Motivation through the design of work: Test of a theory. *Organizational Behavior and Human Performance, 16*(2), 250–279.

Hendrickx, S. (2010). *The adolescent and adult neuro-diversity handbook: Asperger syndrome, ADHD, dyslexia, Dyspraxia and related conditions*. Jessica Kingsley Publishers.

Hennekam, S., & Follmer, K. (2024). Neurodiversity and HRM: a practice-based review and research agenda. *Equality, Diversity and Inclusion, 43*(7), 1119–1129.

Microsoft. (2025). Microsoft neurodiversity hiring program & FAQ. Retrieved 25.01.2025. https://www.microsoft.com/en-us/diversity/inside-microsoft/cross-disability/neurodiversityhiring.

Rollnik-Sadowska, E., & Grabinska, V. (2024). Managing neurodiversity in workplaces: A review and future research agenda for sustainable human resource management. *Sustainability, 2024*(16), 6594.

Szulc, J. M. (2024). Leading with understanding: cultivating positive relationships between neurotypical leaders and neurodivergent employees. *Employee Relations: The International Journal, 46*(9), 97–114.

Trost, A. (2020). *Human resources strategies: Balancing stability and agility in times of digitization*. Springer.

5 Ambidextrous Leadership in the Context of Neurodiversity

5.1 Scope and Definition

Ambidextrous leadership is a recently introduced term by scholars to characterize a special approach to leadership that is mostly used in organizations. It refers to the simultaneous use of explorative and exploitative activities by leaders (Helmold et al., 2023). Ambidextrous leadership is a leadership skill that allows you to flexibly switch between opposing leadership styles depending on what the current situation requires. This type of leadership leverages the strengths of both results-oriented and people-focused approaches to successfully manage and adapt teams (Duwe, 2022). It is a valuable tool for effectively navigating and achieving peak performance in today's dynamic business world. Ambidexterity or Multiple Hands Leadership is becoming increasingly important in everyday management work. This refers to the ability to reconcile different ways of acting and apparent contradictions such as hierarchy versus self-organization and to use them for transformation processes. The classic manager ensures that his area functions reliably. Classic or exploitative leadership is therefore concerned with maintaining and optimizing existing structures (March, 1991). Managers set goals, conduct performance checks, give systematic feedback and reward or punish depending on the degree to which goals are achieved. The role of the agile manager, on the other hand, is a new one. The agile manager ensures that his team has the best possible conditions to complete its tasks in the most self-organized way possible. This form of explorative leadership relies on role models, inspiring motivation, intellectual stimulation, and the individual development of individual team members. In their pure form, these two leadership approaches are rare. In the current transformation processes, managers move between the two poles of "totally exploitative" or "totally explorative." In concrete terms, this means that managers are currently required to use both leadership approaches in parallel, depending on the situation. In this context, ambidexterity describes the ability of managers to switch back and forth between these two

M. Helmold, M. Martensen, *Leadership in Neurodiversity and Neurodivergence*, Diversity and Inclusion Research, https://doi.org/10.1007/978-3-032-09343-1_5

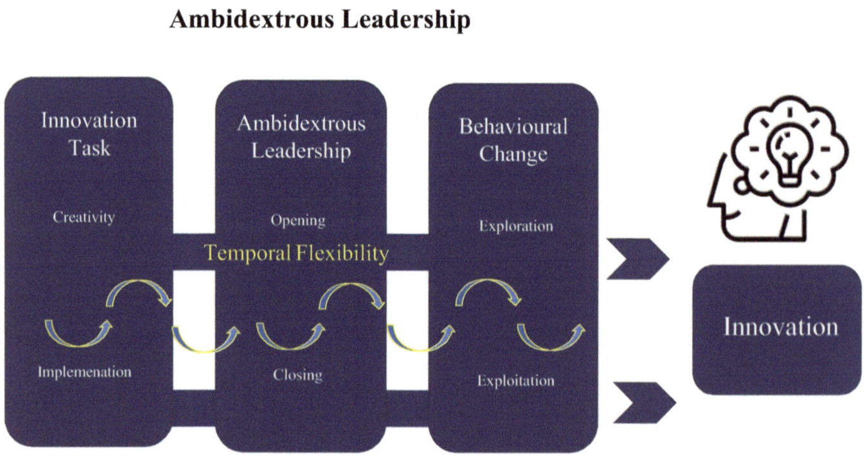

Fig. 5.1 Ambidextrous leadership process. Source: Author

different, sometimes even contradictory, approaches (Kassotaki et al., 2019). Depending on the situation, either approach is implemented (O'Reill & Tushmann, 2004). This can mean, for example, that, on the one hand, the manager simply sets goals and leaves implementation entirely to the team (exploratory) and, on the other hand, has to ensure that routines are adhered to and rules are followed (exploitative). The existing contradictions are obvious. Ambidextrous managers must be able to deal with contradictions and maintain a good balance, as well as have a high degree of empathy for their employees (Ferrary, 2011). Figure 5.1 shows the ambidextrous leadership process. The concept suggests the simultaneous use of explorative and exploitative activities by leaders. Exploration refers to search, risk taking, experimentation, and innovation in organizations, whereas exploitation has to do with refinement, efficiency, implementation, and execution Raisch and Birkinshaw (2008). Successful ambidextrous leaders must be able to achieve the appropriate mix of explorative and exploitative activities, unique for each organization, that will lead them to innovations and significant performance outcomes.

Executive leadership can use the ambidextrous organizational model to create segregated business units for exploring and developing breakthrough innovations (products, services, and processes) while at the same time keeping existing business units intact. Project teams within the new venture are encouraged to form their own processes, structures, and cultures, but they are still connected to the rest of the organization through executive sponsors who ensure that no organizational conflicts or competition for resources threatens the viability of the venture.

5.2 Ambidextrous Leadership on the Microlevel

On the individual level of analysis, scholars have mostly used paradox theory to explain how individuals manage the competing demands of exploration and exploitation. Paradoxical leadership refers to seemingly competing, yet interrelated, behaviors to meet structural and follower demands simultaneously and over time. Managing complex business models effectively depends on leadership that can make dynamic decisions, build commitment to specific visions and goals, learn actively at multiple levels, and engage conflict.

On the group level of analysis, scholars have examined how team members affect leadership actions in the promotion of ambidexterity. Many researchers have focused their attention on the heterogeneity of both board of directors and top management teams. Research has shown that when the functional background heterogeneity of boards of directors is included into the strategy making, then their knowledge resources can contribute to the relative exploration orientation of an organization. In the same vein, top management teams heterogeneity moderates the impact of the internal and external advice seeking on the exploratory innovation.

5.3 Ambidextrous Leadership on the Macrolevel

The roles and behaviors of effective top managers differ considerably from those of middle managers. This can be attributed to the fact that middle managers have different priorities from those of senior executives. When ambidexterity penetrates across levels, explorative and exploitative knowledge sharing of middle managers is especially important for the long-term firm performance. For example, top-down knowledge inflows of managers from the higher hierarchical levels positively relate to middle managers' exploitative activities, while bottom-up and horizontal knowledge inflows from lower levels and peers positively relate to their explorative activities.

On the employee level of analysis, both psychological factors and paradoxical leadership of group managers predict employees' ambidextrous behavior. For instance, ex-ante incentives (incentives based on past performance) and ex-post incentives (incentives based on future performance) affect productivity, motivation, and performance of employees, while making them feel the sense of stretch essential in building an ambidextrous organization.

5.4 Organizational Ambidexterity

Organizational ambidexterity is an organization's ability to simultaneously excel at both exploitation (making current operations more efficient) and exploration (developing new products and markets for future growth). This is crucial for companies to stay competitive by improving what they already do while also innovating for future change (Nieto-Rodriguez, 2014). The most-accepted definition of ambidexterity is a balance between explorations and exploitation: organizations capable of exploiting their existing competencies while simultaneously exploring new opportunities. James March refers to this as the exploration of new possibilities and the exploitation of old certainties. Exploitation includes such things as choice, refinement, production, selection, execution efficiency, and implementation. While exploration encompasses knowledge creation and analysis of future opportunities. Organizations that engage in exploration to the exclusion of exploitation are likely to find that they suffer the costs of experimentation but without gaining many of its benefits. These companies exhibit too many undeveloped new ideas and often too little distinctive competence. A well-known example of too much emphasis on exploration is Ericsson, the telecom giant that led the development last century of the global system for mobile communications. At its peak, its R&D organization employed 30,000 people in 100 technology centers and with considerable duplication of work. Despite its strong focus on exploration, the company's results went into a steep decline. Ericsson laid off around 60,000 employees and closed most of its technology centers to put focus back on exploitation in order to return its businesses to profitability. Conversely, organizations that engage in exploitation to the exclusion of exploration are likely to find themselves trapped in stable equilibrium, going nowhere fast but efficiently. Maintaining an appropriate balance between exploration and exploitation is a primary factor in the prosperity of any corporate system.

5.5 Ambidextrous Leadership and Neurodiversity

Diagnosis of neurodiversity has increased rapidly in recent years. Current estimates reveal 15–20% of the world's population exhibiting some form of neurodivergence. This creates both opportunities and challenges for organizations who want to better support neurodiverse colleagues as well as making full use of the talents of those individuals.

Fundamentally, our success will come down to the effectiveness of how people are managed and led. Unfortunately, many managers have not received an education in how to adapt their style for neurodiverse team members. Importantly, neurotypical people often find it hard to understand the way neurodiverse people process information and this can lead to poorer outcomes for the individual. Neurodiverse and ambidextrous leadership can help companies become more creative, innovative, and diverse. By promoting neurodiversity at the leadership level, organizations can gain a competitive advantage while fostering an inclusive corporate culture. The benefits of neurodiversity in the context of ambidextrous leadership are as follows:

- Creativity and innovation: Neurodiverse individuals often bring unique ways of thinking and problem-solving that can lead to innovative ideas.
- Improved problem-solving: They can analyze and solve complex problems in ways that may be difficult for others to understand.
- Diversified teams: By promoting neurodiversity in leadership, companies create diverse teams that are better able to respond to the needs of a broad group of customers.

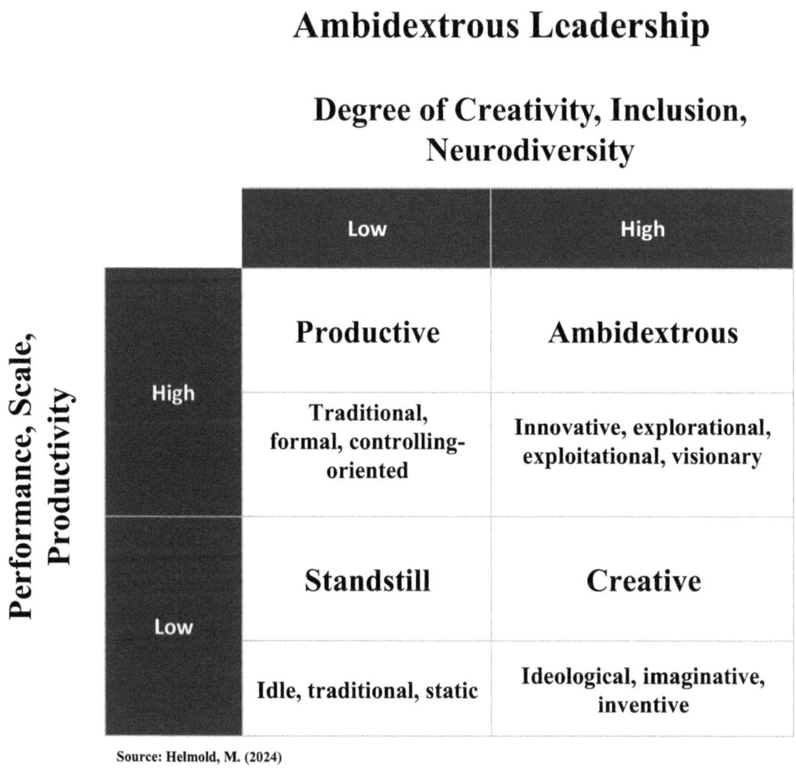

Fig. 5.2 Ambidextrous leadership and neurodiversity. Source: Author

Companies can be considered ambidextrous when they continuously solve the trade-off between being fast, creative, and scale-driven/productive. As such, they establish an equilibrium characterized by a strong emphasis on both dimensions. The vast majority of companies focus on one or the other dimension; those that manage to excel in both areas are rare.

As shown in Fig. 5.2, businesses with an emphasis on speed/creativity possess strong capabilities that enable anticipation, innovation, and adaptation. These translate into corporate attributes such as insight and foresight, inspiration and passion, and trial and error.

5.6 Case Study: Cisco Applying the Ambidextrous Leadership Style

Cisco is seen as a role model and benchmark for ambidextrous integration because it effectively balances innovation and operational efficiency, which is crucial for long-term success, especially in the technology sector. This balance is achieved through a combination of strategic leadership, a focus on both internal development and external acquisition (Acquisition and Development or A&D), and the use of technology to manage both new ideas and established operations. Cisco's success demonstrates how companies can thrive by embracing both exploitation (efficiently running the current business) and exploration (developing new opportunities) (Dong et al., 2023). The leadership strategy encounters five pillars: Collaborate, Learn, Execute, Accelerate, and Disrupt (Cisco, 2024). Each pillar encompasses leadership expectations or competencies that together constitute the skill set of an effective leader.

- **Collaborate**: Working across boundaries, building teams, managing conflict, earning trust, and recognizing good performance
- **Learn**: Developing personal skills and coaching others
- **Execute**: Solving problems, making decisions, delegating, giving feedback, and demonstrating passion for the work
- **Accelerate**: Communicating goals and building capabilities
- **Disrupt**: Envisioning opportunities, innovating, taking risks, and leading change

In this sense, they are expected to possess very high dynamism and potential for growth. As they usually have a multinational presence, they have become the most significant contributors to the growth of the economies of the host countries in which they operate as they commit very high investments in capital and infrastructure and receive enormous revenues. In order to support their operations, these companies extensively use software tools that are provided through internet networking to enhance the communication between their departments that, in several cases, are spread across different continents. The Cisco Leadership Series is a set of four programs that offer a high-impact learning experience focused on building foundational leadership capabilities. It specifically targets managers and other leaders within the company who demonstrate a track record of high performance. Leadership ambidexterity in Cisco refers to the capability of leaders to simultaneously manage current operations while exploring new avenues for growth and innovation. Imagine a leader who can skillfully meet quarterly targets while also fostering an environment of innovative thinking for future projects. To develop an ambidextrous organization, leadership must possess the ability to attend to already existing products, services, business models, and processes while simultaneously supporting the innovations that will drive the organization's future. Ambidextrous design and management ultimately means continuing to oversee the core business (exploitation) while concurrently exploring opportunities to innovate and adapt for the future. This dual capability allows an organization to achieve both operational efficiency in its existing markets and adaptability to changing demands.

Key aspects include:

- **Exploitation (Core Business)**: This involves optimizing existing processes, products, and resources to maximize efficiency, quality, and short-term profitability. It relies on established methods and consistent execution.
- **Exploration (Innovation)**: This refers to searching for new ideas, experimenting with unproven concepts, technologies, and business models, and entering new markets. It requires flexibility, risk-taking, and a tolerance for failure.

The challenge of ambidextrous management lies in balancing these two fundamentally different activities, which often require different organizational structures, cultures, and leadership styles, without letting one compromise the other. The goal is to ensure the company remains competitive in the present while positioning itself for long-term success and growth.

References

Cisco (2024). Cisco. our employees. Five themes for effective leadership. Retrieved December 26, 2024. https://www.cisco.com/#tabs-7edb32179e-item-d3e62c5c0c-tab.

Dong, X., et al. (2023). *Cisco. Integration of innovation and operation*. Springer.

Duwe, J. (2022). *Specialized organizations and ambidextrous clusters in the open innovation paradigm Ambidextrous Leadership. How leaders unlock innovation through ambidexterity*. Springer.

Ferrary, M. (2011). Specialized organizations and ambidextrous clusters in the open innovation paradigm. *European Management Journal, 29*(3), 181–192.

Helmold, M. et al. (2023). New Work, Neues Arbeiten virtuell und in Präsenz. Konzepte und Werkzeuge zu innovativer, agiler und moderner Führung. Springer .

Kassotaki, et al. (2019). Ambidexterity penetration across organizational levels in an aerospace and defense organization. *Long Range Planning, 52*(3), 366–385.

March, J. G. (1991). Exploration and exploitation in organizational learning. *Organization Science, 2*(1), 71–87.

Nieto-Rodriguez, A. (2014). Organisational ambidexterity. Understanding an ambidextrous organisation is one thing, making it a reality is another. London Business School Retrieved December 26, 2024. https://www.london.edu/think/organisational-ambidexterity.

O'Reilly, C. A., & Tushmann, M. L. (2004). The ambidextrous organization. *Harvard Business Review, 82*(4), 74–81.

Raisch, S., & Birkinshaw, J. (2008). Organizational ambidexterity: Antecedents, outcomes and moderators. *Journal of Management, 34*(3), 375–409.

6 Self-Leadership and Neurodiversity

6.1 Scope and Definition

The working world is becoming increasingly complex and constantly presents us with new challenges. In order to be able to overcome these, the ability to lead oneself is becoming increasingly important. This is especially true for managers because if you are not able to lead yourself, you cannot lead anyone else. This makes self-leadership a core competency for managers (Helmold et al., 2023). In a working world in which managers and employees are constantly confronted with new and increasingly complex requirements, self-leadership is a key skill (Covey & Franz, 2024). Successful self-leadership helps to constructively change and positively influence your own feelings and emotions. You get to know yourself better and better, resolve "blind spots" and thus create the basis for being able to lead successfully and live up to your role model function for employees. Self-leadership is often used synonymously with self-organization, self-management, self-control, self-regulation, or self-reflection. These terms often go hand in hand, but are not completely the same. Self-leadership means being able to direct thoughts, emotions, and behaviors in a desired direction through self-reflection, self-evaluation, and self-motivation. This can lead to an increase in performance and make it easier to achieve set goals. However, self-leadership should not only be associated with increased performance and better performance. It is also about critically questioning yourself and evaluating your own behavior from time to time and adapting it if necessary.

6.2 Advantages of Self-Leadership in Neurodiverse Teams

Successful self-leadership brings numerous advantages for neurodivergent and diverse teams. If practiced successfully, self-leadership enables you to proactively shape your professional and personal development, define goals precisely, and increase intrinsic

M. Helmold, M. Martensen, *Leadership in Neurodiversity and Neurodivergence*, Diversity and Inclusion Research, https://doi.org/10.1007/978-3-032-09343-1_6

motivation so that it is easier to achieve these goals. Those who successfully lead themselves are clear about their own priorities and know that they alone are responsible for taking the appropriate steps if something is to change in the current situation.

As many advantages as self-leadership brings with it, the challenges on the way to learning this skill are just as great. The problem: In order to be able to lead themselves, managers must first look inward. However, many of them find this difficult because they are more used to observing and judging others in their behavior and actions. Instead, they must learn to focus on their own values and norms. What is important to them? What do they want to stand for in their role model function as a manager? To ensure that this change and thus self-leadership succeed, we present you with three important building blocks of successful self-leadership.

6.3 3S-Concept for Self-Leadership: Self-Awareness, Self-Regulation, and Self-Responsibility

6.3.1 Self-Awareness

Self-Awareness In relation to self-leadership, self-awareness is a significant element. Self-awareness can be described as self-knowledge or self-perception (Helmold et al., 2023). This means the ability to perceive and reflect on oneself and one's own strengths and weaknesses. Self-awareness thus forms the basis for emotional intelligence, which is an indispensable skill for both managers and employees today (Furtner & Baldegger, 2024). Those who are clear about their own emotions, behaviors, and driving forces are more likely to be able to prevent destructive or regressive actions in extreme situations. However, developing self-awareness is a process that takes time and also requires a certain amount of self-forgiveness. It is important to take a step back, observe, and reflect on yourself. In this way, one's own strengths and weaknesses can be uncovered and appropriate change processes can be initiated. How can this be achieved? Firstly, by getting feedback on your own leadership qualities. The basic requirement: It has to be honest and open feedback. Secondly, it is important to proceed with the journaling of feelings on a regular basis (Helmold, 2022). Even if the idea of keeping such a diary may initially meet with inner resistance in some people, one cannot progress on the path to self-knowledge without this step. Writing down daily emotions and weighting them by intensity is a form of journaling that builds self-awareness by tracking patterns, processing feelings, and understanding your inner world. This practice, combined with consciously finding new communication methods, can improve emotional regulation, stress management, and relationships. This will help leaders get to know the inner landscape better and better understand personal patterns. Another helpful exercise that helps with self-reflection is to ask yourself the following questions:

- What am I doing, why am I doing it?
- What effect does this have on others?
- What effects does this have on me?
- What are good, what are bad emotions?
- Why did I react in a way, positively or negatively?

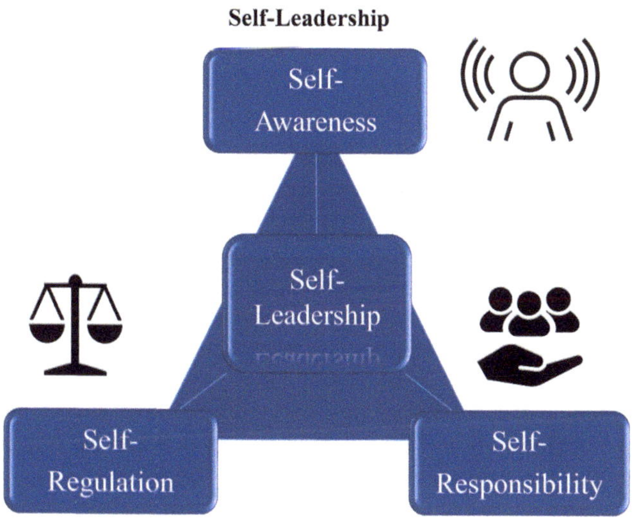

Fig. 6.1 Self-leadership elements

Tip for Leaders Leaders should write down the answers to these essential questions on a daily and weekly basis so that they can look at them again at any time and adapt them if necessary. The transparent analysis of good and bad emotions will help them get a step back and to reflect better (Fig. 6.1).

6.3.2 Self-Responsibility

Self-responsibility is the principle that individuals are accountable for their own choices, behaviors, actions, and the outcomes that follow. It involves acknowledging your role in your life's events rather than blaming external factors or other people.

Leaders who reflect and act responsibly show reliability. This makes it easier for them to gain the trust of their employees, which in turn is the basic prerequisite for excellent performance and successful collaboration.

The first step in learning self-responsibility is to change the inner attitude. To do this, leaders need to not always see themselves as prisoners of circumstances, but instead accept that ultimately only they are responsible for whether they are satisfied, happy, and successful. This also includes admitting mistakes, not always blaming others and proactively looking for solutions to problems instead of hoping that someone else will take care of them. This will make it easier and easier to automatically take responsibility for yourself.

Tip for Leaders Self-responsibility is a cornerstone of effective leadership, requiring a fundamental shift in inner attitude that directly translates into conscious behavior and actions. This elaboration involves moving away from a mindset of victimhood or blame and fully owning one's choices and their consequences for self-responsive leaders.

6.3.3 Self-Regulation

Self-regulation is a process involving different mechanisms, such as attention control, emotion regulation, impulse control, and goal setting. Being able to self-regulate, control emotions and behaviors and resist impulses can allow you to become a stronger leader. Great leaders are able to effectively self-regulate. Self-regulation is the ability to control or redirect disruptive impulses and moods; the propensity to suspend judgment—to think before acting. Self-regulation includes sensory regulation, emotional regulation, and cognitive regulation (Helmold et al., 2023). The primary reason to focus on self-regulation is that it generally promotes a more positive response, interaction, and answer in any given situation.

Tip for Leaders Leaders must control and redirect impulses and moods. It is a statement and demonstration of calmness in any given situation. It is being able to control emotions and responses to situations and other people.

6.4 Neurodivergent People as Leaders

Neurodiverse individuals bring unique strengths that can drive innovation and enhance team performance, making them great candidates for leadership positions. Neurodivergent leaders offer a different perspective and thought process that can open up new opportunities for enterprises and organizations. Neurodivergent leadership may foster innovation, agility, and prosperity (Institute of Leadership, 2024). Neurodiverse leaders often excel in creativity, have great attention to detail, and offer fresh approaches to challenges others may not see. Leaders with ADHD may also have higher energy levels, which can be great for generating a positive atmosphere at work, and boosting morale within teams. There are several examples of neurodivergent people as successful leaders, e.g. entrepreneur Richard Branson, environmentalist Charles Darwin, painter Vincent Van Gogh, and the composer Mozart (World Economic Forum, 2023).

6.5 Case Study: Self-Leadership in the Virgin Group by Richard Branson

The self-leadership style of Richard Branson embodies a unique blend of charisma, innovation, and agile entrepreneurship. Known globally for steering the Virgin Group to astronomical success, Branson's approach to leadership has intrigued and inspired many. Branson is a strong believer in empowering his employees. He often emphasizes the importance of listening to employees at all levels and encouraging them to contribute ideas (Branson & Jacobs, 2012). This approach creates a sense of responsible ownership and engagement among staff. Branson is known for his willingness to take risks and innovate. He encourages creative thinking and is not afraid to venture into new and untested markets or industries. This aspect of his leadership has been crucial in the diversification of the Virgin brand into various sectors. Unlike many traditional business leaders, Branson adopts a more informal and down-to-earth style. He is often seen as a leader who is easy to talk to and works alongside his employees rather than positioning himself above them (Branson, 2007).

Born in July 1950, Richard Branson has been a charismatic presence in UK entrepreneurship since the early 1970s. After setting up a magazine, he established a mail order import record business. He later founded the Virgin Records and opened a chain of record shops, Virgin Megastores. Branson diversified Virgin into transport with Virgin Atlantic airlines and Virgin Trains (between 1997 and 2019). Further ventures in telecommunications, well-being, and finance followed, and Branson joined the exclusive club of space-age billionaires with the launch of Virgin Galactic in 2004. Branson is considered an entrepreneur, risk-taking, innovative, and agile.

References

Branson, R. (2007). *Screw it, let's do it: Lessons in life and business*. Virgin Books.

Branson, R., & Jacobs, C. (2012). *Geht nicht gibt's nicht!: So wurde Richard Branson zum Überflieger. Seine Erfolgstipps für Ihr (Berufs-) Leben*. Börsenmedien Frankfurt.

Covey, S. R., & Franz, C. (2024). *Self-Leadership: Die 7 Wege zu einem selbstbestimmten und sinnerfüllten Leben (Dein Erfolg)*. Gabal.

Furtner, & Baldegger. (2024). *Self-Leadership und Führung: Theorien, Modelle und praktische Umsetzung* (3rd ed.). Springer.

Helmold, M. (2022). Leadership. Agile, virtuelle und globale Führungskonzepte in Zeiten von neuen Megatrends. Springer .

Helmold, M., et al. (2023). New Work, Neues Arbeiten virtuell und in Präsenz. In *Konzepte und Werkzeuge zu innovativer, agiler und moderner Führung*. Springer.

Institute of Leadership (2024). It's time to embrace neurodiversity. Neurodiversity is the future. Retrieved December 30, 2024. https://leadership.global/resourceLibrary/embracing-neurodiversity.html#:~:text=The%20world%20has%20been%20influenced,Goch%20and%20the%20composer%20Mozart.

World Economic Forum. (2023). Forum institutional. Employers now see neurodiversity as a strength in the workplace. Here's why. Retrieved December 24, 2024. https://www.weforum.org/stories/2023/05/neurodiversity-employers-workers-jobs/.

Equity, Equality, Diversity, and Inclusion at the Workplace

7

7.1 Equity and Equality

Equity recognizes that each person has different circumstances and allocates the exact resources and opportunities needed to reach an equal outcome. Equality means each individual or group of people is given the same resources or opportunities. Equality is about being equal in status, rights, and opportunities, whereas equity is about how we get there through fairness and impartiality (Frones et al., 2020). Equity and equality both recognize that fairness is important, but they differ in how they achieve it. Equality gives everyone the same resources and opportunities, while equity acknowledges that individuals have different circumstances and need different levels of support to achieve an equal outcome.

Equality provides identical resources to everyone, assuming a level playing field. Equity provides tailored resources based on an individual's or group's unique needs to correct for disadvantages.

7.2 Importance of Work Equality

7.2.1 Formal, Social, Individual, and Systematic Equality

Promoting and delivering equality, diversity, and inclusion (EDI) in the workplace is an essential aspect of good people management and leadership. EDI should not only be driven by laws, regulations, or ethical standards, it should also be followed for being more productive (Helmold et al., 2023). High levels of equality, diversity, and inclusion are associated with greater innovation, productivity and performance, talent recruitment and retention, and workforce well-being (Ely & Thomas, 2020).

EDI is about creating working environments and cultures where every individual can feel safe, experience a sense of belonging, and is empowered to achieve their

M. Helmold, M. Martensen, *Leadership in Neurodiversity and Neurodivergence*, Diversity and Inclusion Research, https://doi.org/10.1007/978-3-032-09343-1_7

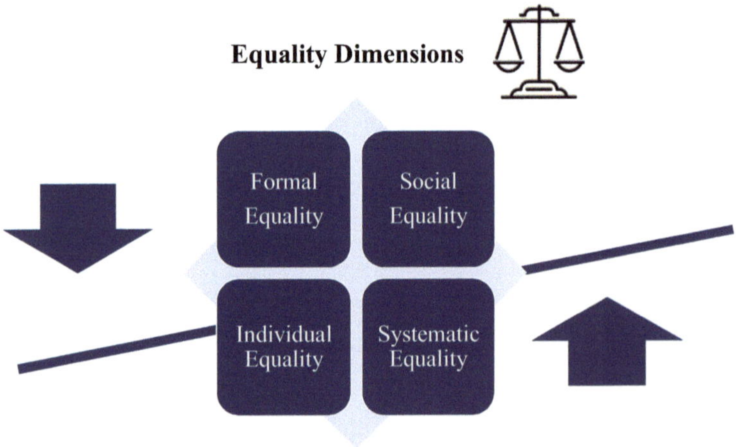

Fig. 7.1 Equality dimension. Source: Author

full potential. Workplace equality refers to being treated fairly and equally in the workplace regardless of gender, race, age, religion, sexual orientation, or any other personal characteristic. Everyone should have equal access to opportunities, resources, and benefits and not be discriminated against or treated differently based on factors unrelated to job performance and skills. Workplace equality also means that diverse perspectives and experiences are valued and leveraged to create a more inclusive and innovative working environment. Achieving workplace equality requires commitment and organizational buy-in to create policies and practices that promote fairness, respect, and equal opportunities for all employees.

Figure 7.1 depicts the elements of formal, informal, individual, and systematic equality in organizations.

7.2.2 Formal Equality

Equality focuses on treating everyone the same, providing equal opportunities and resources to all individuals. At work, formal equality appears as a written set of rules that dictates the equal treatment of all people within an organization. One example happens when positions open to all (qualified) candidates available for hire. This means any person with a fair application has a chance of getting the job. Formal equality is an equality of men and women in the formal sense, that is all are same and are subject to the same rules and regulations in social, political, economic, and other areas of life. The same principles apply to neurodivergent people. Many enterprises have established support mechanisms and reasonable adjustments, which are in place, to remove any barriers or discrimination for neurodivergent people.

Neither discrimination nor privilege is given to any section of people in the society. Further, formal equality does not recognize the special needs of different people who are of different capacities, needs, and interest. Formal equality lays emphasis on the idea that all are given equality to enjoy their life according to their own capacities and expecting none to act on their behalf.

The recognition of invisible disabilities such as specific learning disabilities by the European Union is very important because it allows the rights of people to be asserted in all countries of the Union. The EU-funded NEDBELS project is investigating the legal and social challenges society faces if it is to be more inclusive of those with such neurological disorders, the concept known as neurodiversity. The project was launched to address a perceived gap in communications between legal experts and scientists on the treatment of those with neurodevelopmental disorders—which affects formal training and daily practice in diagnosis, education, and health services for those affected (European Commission, 2024).

7.2.3 Social Equality

Social equality is a state of affairs in which all individuals within society have equal rights, liberties, and status, possibly including civil rights, freedom of expression, autonomy, and equal access to certain public goods and social services. Social equality requires the absence of legally enforced social class or caste boundaries and the absence of discrimination motivated by an inalienable part of an individual's identity. For example, advocates of social equality believe in equality before the law for all individuals regardless of sex, gender, ethnicity, age, sexual orientation, origin, caste or class, income or property, language, religion, convictions, and opinion (Helmold, 2022).

The rights of people with neurodivergent disorders pertain to social equality. Both physical and mental disabilities can prevent individuals from participating in society at an equal level, due to environmental factors as well as stigmas associated with disability (Rieser, 2011). Social equality includes both the treatment of disabilities and the accommodation of people with disabilities to facilitate equal participation in society.

7.2.4 Individual Equality

Individual equality is the state of being equal, especially in status, rights, and opportunities. Equality means each individual or group of people is given the same resources and opportunities, regardless of their circumstances. A rewards and recognition system must take into account individual equity when the compensation of incumbents holding the same job in the company varies based on years of service, performance, skills, and experience. Individual equality measures must be designed to make sure that employees with neurodiverse characteristics or disabilities are not at a disadvantage. The word individual makes clear that equality measures are tailored to the individual abilities of the employee while at the same time making sure that they are in accordance with the specific expectations according to needs and law.

7.2.5 Systematic Equality

Creating systemic equality means ensuring that everyone has access to the resources and tools they need to be successful. Systemic inequality refers to disparities in wealth, resources, opportunities, and privileges that result from discriminatory practices and are deeply ingrained within the structures and institutions of society.

7.3 Legal Aspects of Work Equality

Work equality is integrated in the legal system. The General Act on Equal Treatment (German abbreviation, AGG) has been in force since August 18, 2006. It incorporates four Anti-Discrimination Directives of the EU into German law. The purpose of this Act is to prevent or to stop discrimination on the grounds of race or ethnic origin, gender, religion or belief, disability, age, or sexual orientation. Section 1 of this Act, The General Act on Equal Treatment, governs the claims and legal consequences in the case of discrimination, both in the field of work and also for the sphere of civil law (Antidiskriminierungsstelle, 2024).

Equality before the law, also known as equality under the law, equality in the eyes of the law, legal equality, or legal egalitarianism, is the principle that all people must be equally protected by the law. The principle requires a systematic rule of law that observes due process to provide equal justice, and requires equal protection ensuring that no individual or group of individuals be privileged over others by the law. Also called the principle of isonomy, it arises from various philosophical questions concerning equality, fairness, and justice (Schach, 2023). Equality before the law is one of the basic principles of some definitions of liberalism. It is incompatible with legal slavery.

Article 7 of the Universal Declaration of Human Rights (UDHR) states: “All are equal before the law and are entitled without any discrimination to equal protection of the law.” Thus, it states that everyone must be treated equally under the law regardless of race, gender, color, ethnicity, religion, disability, or other characteristics, without privilege, discrimination, or bias (Helmold, 2022). The general guarantee of equality is provided by most of the world’s national constitutions, but specific implementations of this guarantee vary. For example, while many organizations guarantee equality regardless of race, only a few mention the right to equality regardless of nationality.

7.4 Diversity and Inclusion

In the recent years, it is visible that the focus and awareness of diversity and inclusion are permanently increasing. Society, enterprises, and organizations face many challenges in identifying and truly understanding the unique differences among people. From the #MeToo movement to various head-line scandals, diversity and inclusion have been brought to the forefront of workplace dialogue (Helmold et al., 2023). However, the words “diversity” and “inclusion” are often confused.

Diversity is the presence and acceptance of differences at the workplace, whereas inclusion is the active, intentional, and ongoing engagement with diversity to achieve a culture in which different people can come together to work, feel comfortable and confident to be themselves, and feel valued. Diversity in the workplace refers to the variation in personal, physical, and social characteristics, such as gender, ethnicity, age, and education. Inclusion, on the other hand, pertains to the corporate strategies and procedures organizations implement to integrate everyone, ensuring that their differences coexist beneficially (Helmold et al., 2023).

7.5 Fostering Diversity at the Workplace

The concept of diversity has its origins in the US civil rights movement, which fought against racism against people of color. Diversity thus initially stood for the creation of equal opportunities for groups who are disadvantaged according to certain characteristics. In the USA, this resulted in the Anti-Discrimination Act and the Affirmative Action to promote disadvantaged groups according to the criteria of race, gender, skin color, ethnic origin, age, disability, or religion. Yes, the Civil Rights Movement of the USA significantly influenced the development of social movements for Native Americans and other groups facing discrimination, demonstrating the power of nonviolent activism, legal challenges, and public awareness to fight for equality (Helmold et al., 2023). The European Union has also used the concept as a model since the late 1990s. Since 2006, the aspects of diversity have been taken into account in German legislation in the General Equal Treatment Act and are intended to protect people from these categories from discrimination.

However, research into and awareness of these issues were not tied to the concept of diversity. In social psychology, heterogeneity had been the central research-leading term for the dimensions examined since the late 1940s. For decades, social psychology has been researching the mechanisms of the emergence and effects of social categorization and stereotyping, social comparisons, identity formation in social groups, and intergroup processes. The effects of belonging to social groups have been extensively researched, especially with regard to educational and performance behavior, processes of social advancement and decline, crime and gang formation, health and working life. In this respect, the results of group-related research were of extraordinary importance for education and social medicine.

Diversity management as a method of corporate human resource management and workplace management. It aims to use the diversity of employees constructively, efficiently, and profitably. A well-known representative of diversity management in Germany is the economic policy initiative Charter of Diversity. The Diversity Charter is a voluntary commitment published in 2006 and an association under the patronage of the Federal Chancellor, which advocates a working environment free of prejudice. By signing the Diversity Charter, employers declare that they will create or promote equal opportunities for their employees. In 2018, there were 3000 signatories (as of September 2018). In addition to large corporations, small and medium-sized enterprises (SMEs), academic and social institutions, and government authorities are fostering the diversity approach within their organisations (Charta der Vielfalt, 2020).

Employers have to observe the ban on discrimination when publishing job advertisements and also during the application procedure. However, the employees—women and men—are also entitled to be protected against discrimination in existing employment relationships. They may lodge a claim for compensation or damages and file a complaint about discrimination with their employers. For this purpose, a complaints unit has to be established at all companies and the employees have to be duly notified about its existence. Employers have to ensure that discrimination does not take place. In addition, they are obliged to take measures against employees who discriminate against other colleagues. The possible measures for this purpose range from a transfer to another post over a reprimand up to a dismissal (Helmold et al., 2023).

But there are also exceptions to the ban on discrimination: For example, a denominational nursery school is free to require a certain religious affiliation as a condition for employing a nursery school teacher. In the field of civil law, the protection granted in the General Act on Equal Treatment comprises bulk business and insurances under private law. Bulk business is concluded irrespective of the person involved and in a large number of cases. These include everyday matters such as shopping or going to restaurants. The protection by law also takes effect in the case of insurance under private law, which has been taken out later than December 22, 2007. However, where insurance contracts have been modified subsequent to the introduction of the General Act on Equal Treatment (August 18, 2006), these modifications might—depending on the respective circumstances—infringe upon the provisions of the General Act on Equal Treatment. In cases of infringements of the General Act on Equal Treatment, persons discriminated against may assert claims to compensation or damages.

Workplace diversity has become complex due to the increasing scope of what constitutes diversity (including age, religion, and cognitive style) and the challenges of navigating diverse perspectives, which can lead to communication barriers, unconscious bias, and resistance to change. Organizations face difficulties in fostering a truly inclusive environment where diverse employees feel valued and respected, which requires more than simply hiring a diverse workforce. There has been increasing awareness of how diversity could be management and this has extended beyond the demographic characteristics of race, gender, age, and education. Countries around the world are gaining larger populations and becoming diverse.

Diversity has meanwhile become a crucial topic to strive for excellence and meet customer expectations in industry and academia. Although diversity is largely promoted in larger multinational companies (MNC), it is important to understand that diversity must be integrated in any organization, enterprise, or institution. MNC widely promote their diversity activities through social and online networks as a way to promote the employer and company branding (Helmold, 2022). Specific diversity programs foster this trend.

Diversity has enabled up-surging of creativity, innovation, and improved problem-solving, which in turn boost the effectiveness of the organization. An organization that has good diversity workplace will generate an increase to their market share if they take care of their diverse customers. However, the fact remains that diversity on its own won't improve business performance (Mccuiston et al., 2004).

Discrimination is becoming widespread in the workplace and is harmful to organization. Any modern business goals should always consist of abolishing all kinds of discrimination around their work environment and create a positive environment that cuddles diversity. It is vital to get a clear vision that discrimination in the work place is not narrowed to just a selected group of individuals but rather it affects any person regardless of his or her race, religion, culture, or sexual orientation. Therefore, all organizations have every reason to institute effective diversity management practice that yields the benefit of a diverse workforce (Helmold, 2022).

Companies must enforce diversity because their diversity will improve the workplace by introducing new talents, interest, and viewpoints. Hence, it is possible to achieve organizational excellence and customer satisfaction. Any organization that fails to welcome cultural diversity in its system and that did not take swift action in abolishing discrimination and injustice will horribly affect their employees and clients. Therefore companies have to thus redefine their missions, management practice, strategies, cultures, markets, customers, and stakeholders. Eventually, the main strategy is to design, develop, and maintain diversity in the workplace and create a system that enables the employees to get attached to the corporation.

Figure 7.1 depicts the elements of diversity in terms of age, gender, ethnicity, race, physical abilities and characteristics, religious and spiritual beliefs, abilities, income, education, marital status, parental status, location, personality, or political beliefs (Fig. 7.2).

Diversity and inclusion must be integrated into the company's mission, vision, and values. Many companies therefore introduce diversity programs and activities. These activities will help organizations reach talent in traditionally undertapped demographics (Helmold, 2022).

They can also encourage innovation, according to a Forbes Insight report; and they can make a company a more desirable option to candidates. Two-thirds of respondents in a recent survey from career community site Glassdoor said that they viewed diversity as an important factor when assessing a job (Helmold, 2022).

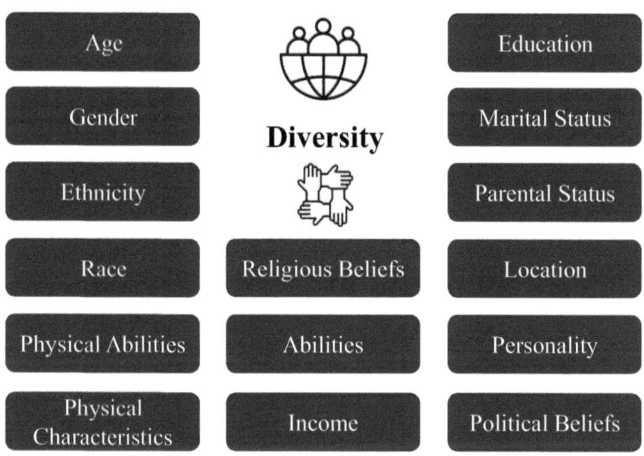

Fig. 7.2 Elements of diversity. Source: Author's Source

However, simply implementing a diversity program, without careful attention to its composition and desired effect, won't significantly improve an organization's organizational change efforts. After years of diversity and inclusion program implementation and research, it is clear that several key factors can potentially make or break a workplace diversity and inclusion program. Although the specifics may vary from organization to organization, for a diversity program to be truly successful, employers need to include several structured elements (Talent Intelligence, 2015).

Figure 7.3 shows the need for diversity cycle with the needs analysis, a training program, the career and diversity-focused leadership, the diversity-specific objectives, and the reviews and calibration. The needs analysis is a systematic process of identifying and prioritizing the gaps between your current and desired state of diversity and inclusion. The needs assessments are designed to provide an overview of any challenges employees might face at work, as a result of neurodiversity or disability. Based on the results, a training and implementation program will be designed. The following steps must be taken:

A Thorough Assessment and Needs Analysis Without a proper analysis and roadmap, according to the Society for Human Resource Management (SHRM), companies may not be able to need to determine what needs to change. An assessment can help your organization accurately identify what elements its diversity and inclusion program should include. The SHRM suggests potentially using focus groups and employee surveys to gauge current employee sentiment and obtain ideas for possible solutions.

Fig. 7.3 Diversity needs analysis. Source: Author

A Solid Training Program Employees—particularly managers who will oversee diversity programs—need to be given a solid framework to work with. A diversity training program can also help convey to your staff, at all levels, that your organization is placing a serious emphasis on diversity and inclusion. Workforce magazine recommends focusing diversity training programs on clear metrics, experiential learning, and sharing an individual benefit for all involved parties to help make diversity program training efforts successful.

Career-Focused Leadership and Guidance Fostering growth from within can help enhance employee retention and provide additional support for diversity efforts. Mentoring programs, according to Scientific American, can be one of the most successful ways to increase the amount of white and black female and Latino and Asian female and male managers at an organization—potentially by nearly 40 percent.

Diversity-Specific Objectives and Targets A general stance to increase diversity is likely far too vague to be very successful. Companies that create clearly outlined goals have a better chance of developing effective programs and seeing real change. Take, for example, Shell. The global energy and petrochemical company's diversity program goals include increasing the proportion of women in senior management roles to at least 20 percent and placing local residents in more than half of the senior management positions in countries it operates in.

Periodic Reviews Goals are only effective if you're able to determine if you're meeting them. An organization's management team and CEO, according to SHRM, should regularly gauge its diversity and inclusion program's success, based on its predetermined goals. If an element isn't working, you'll then be able to tweak the program as needed to see greater results. If your organization hasn't measured its diversity and inclusion status recently, that should be your first step in either creating a new diversity program or revising your current one.

7.6 Case Study: Diversity in Deutsche Telekom

Deutsche Telekom is one of the world's leading integrated telecommunications companies, with more than 252 million mobile customers, 25 million fixed-network lines, and 22 million broadband lines. Deutsche Telekom is present in more than 50 countries. With a staff of some 199,562 (Dec 31, 2023) throughout the world, they generated a revenue of 112 billion Euros in the 2023 financial year (Deutsche Telekom, 2024).

The Deutsche Telekom is committed to ensuring that everyone is offered the same opportunities, regardless of their personal circumstances (Deutsche Telekom, 2024). Their goal is the non-discriminatory inclusion of everyone. A company like Deutsche Telekom thrives on the diversity of its employees. But diversity as a strategy or diversity management means more: a comprehensive management concept that is based on mutual recognition and appreciation and uses the diversity of employees strategically—in the interest of economic success. The customer and business relationships that all strategies are geared toward are just as diverse. Diversity is closely aligned with operational processes, the needs and conditions of the business areas in the group, and the special situation of Deutsche Telekom on the global market. In other words, diversity as a management concept means recognizing and acknowledging individuality. It is about valuing every employee and customer—regardless of, for example, gender, age, disability, ethnic origin, religious beliefs, or sexual orientation (Pagel & Mauz, 2008).

References

Antidiskriminierungsstelle. (2024). Allgemeine Gleichbehandlungsgesetz (AGG). Retrieved 30.12.2024. https://www.antidiskriminierungsstelle.de/SharedDocs/downloads/DE/publikationen/AGG/agg_gleichbehandlungsgesetz.pdf?__blob=publicationFile.

Charta der Vielfalt. (2020). www. Charta-der-vielfalt. Retrieved August 21, 2020. https://www.charta-der-vielfalt.de/.

Deutsche Telekom. (2024). Deutsche Telekom is living diversity. Retrieved December 30, 2024. https://www.telekom.com/en/company/human-resources/diversity#:~:text=Deutsche%20Telekom%20is%20living%20diversity,regardless%20of%20their%20personal%20circumstances.&text=Our%20goal%20is%20the%20non,have%20introduced%20the%20first.

Ely, R. J., & Thomas, D. A. (2020). Diversity and inclusion. Getting serious about diversity: Enough already with the business case. *Harvard Business Review.* Retrieved 30.12.2024. https://hbr.org/2020/11/getting-serious-about-diversity-enough-already-with-the-business-case.

European Commission. (2024). European Commission. Neurodiversity—understanding the legal and social impacts. Retrieved December 30, 2024. https://projects.research-and-innovation.ec.europa.eu/en/projects/success-stories/all/neurodiversity-understanding-legal-and-social-impacts.

Frones, T. J., et al. (2020). *Equity, equality and diversity in the Nordic model of education.* Springer.

Helmold, M. (2022). *Leadership. Agile, virtuelle und globale Führungskonzepte in Zeiten von neuen Megatrends.* Springer.

Helmold, M. et al. (2023). New Work, Neues Arbeiten virtuell und in Präsenz. Konzepte und Werkzeuge zu innovativer, agiler und moderner Führung. Springer .

Mccuiston, V. E., et al. (2004). Leading the diverse workforce: Profit, prospects and progress. *Leadership & Organization Development Journal, 25*(1), 73–92.

Pagel M. and Mauz, B. (2008). Praxisbeispiel Deutsche Telekom: Diversity als Strategie. pp 93–96. In: Krell, G. (2008). Chancengleichheit durch Personalpolitik. Gleichstellung von Frauen und Männern in Unternehmen und Verwaltungen. Rechtliche Regelungen—Problemanalysen—Lösungen. Springer .

Rieser, R. (2011). The struggle for disability equality. In M. Cole (Ed.), *Education, equality and human rights: Issues of gender, 'Race', sexuality, disability and social class* (3rd ed.). Taylor & Francis.

Schach, A. (2023). *Diversity & Inclusion in Strategie und Kommunikation: Vielfalt in Konzeption, Kultur und Sprache im Unternehmen.* Springer.

Talent Intelligence. (2015). The top 5 elements your diversity program should include. Posted by Talent Intelligence. Retrieved August 21, 2020. https://www.talentintelligence.com/blog/the-top-5-elements-your-diversity-program-should-include.

Stress and Neurodiversity 8

8.1 Scope and Definition

Stress is the body's response to a challenging or demanding situation. When someone feels stressed, the body releases certain hormones. The hormones are chemical signals that the body uses to tell the body systems what to do. The hormones the body releases when someone is stressed get them ready to meet the challenge or demand in this dangerous environment. During the stress response, the body gets ready to flee or fight by increasing your heart rate, breathing rate, and blood pressure. Stress is a physical and mental response to perceived threats or challenges. Stress can have positive or negative effects on a person's well-being, depending on the type, intensity, duration, and coping strategies of the individual (Adamsen, 2023). However, too much pressure can become stressful and affect our happiness and quality of life, leaving us feeling physically and mentally unwell.

8.2 Stress Symptoms

Stress can cause many different symptoms. It might affect how you feel physically, mentally, and also how you behave. Sometimes pressure can motivate and make us more productive.

8.2.1 Emotional Stress Symptoms

Mental symptoms of emotional stress include the following:

- Feeling more emotional than usual, especially feeling grumpy, teary, or angry
- Feeling anxious, overwhelmed, nervous, or on edge

M. Helmold, M. Martensen, *Leadership in Neurodiversity and Neurodivergence*, Diversity and Inclusion Research, https://doi.org/10.1007/978-3-032-09343-1_8

- Feeling sad or depressed
- Feeling restless
- Trouble keeping track of or remembering things
- Trouble getting your work done, solving problems, making decisions, or concentrating

8.2.2 Physical Stress Symptoms

Symptoms of stress that you might feel in your body include the following:

- Headaches
- Dizziness
- Clenching your jaw and grinding your teeth
- Shoulder, neck, or back pain; general body aches, pains, and tense muscles
- Chest pain, increased heart rate, heaviness in your chest
- Shortness of breath
- Feeling more tired than usual (fatigue)
- Sleeping more or less than usual
- Upset stomach, including diarrhea, constipation, and nausea
- Loss of sexual desire and/or ability
- Getting sick more easily, such as getting colds and infections often

8.2.3 Respiratory Stress Symptoms

This is when you aren't getting enough oxygen or are having to work really hard to breathe. If you or a loved one has symptoms of respiratory distress, you need to call 911 and get to the ER as soon as possible. Signs include the following:

- Breathing faster than usual.
- Color changes of your skin, mouth, lips, or fingernails. A blue color around your mouth, lips, or fingernails usually shows you aren't getting enough oxygen. Your skin may also look pale or gray.
- Grunting when you breathe out.
- A whistling with each breath (wheezing).
- Nose flaring.
- Chest sinking below your neck or under your breastbone with each breath (retractions).
- Increased sweating, especially cold, clammy skin on your forehead.
- Leaning forward while sitting to help take deeper breaths.

8.2.4 Cognitive Stress Symptoms

Symptoms of stress that affect your mental performance include the following:

- Trouble keeping track of or remembering things
- Trouble getting your work done, solving problems, making decisions, or concentrating
- Feeling less commitment to your work
- Lack of motivation
- Negative thinking

8.2.5 Behavioral Stress Symptoms

Symptoms of behavioral stress include the following:

- Changes in your eating habits; losing or gaining weight
- Procrastinating and avoiding responsibilities
- Using alcohol, tobacco, or drugs to feel better
- Avoiding your friends and family; isolating yourself from others
- Failing to meet your deadlines
- Increased absences at school or work
- Doing your work more slowly
- Exercising less often

8.2.6 Symptoms of Chronic Stress

Chronic stress is when you experience stress over an extended time. This can have negative effects on your body and your mental state, and it can increase your risk of cardiovascular disease, anxiety, and depression.

In general, the symptoms of chronic stress are the same as those for shorter term stress. You may not have all these symptoms, but if you have more than three symptoms and they last for a few weeks, you may have chronic stress. Potential symptoms to look for include the following:

- Aches and pains
- Changes in your sleeping patterns, such as insomnia or sleepiness
- Changes in your social behavior, such as avoiding other people
- Changes in your emotional response to others
- Emotional withdrawal
- Low energy, fatigue
- Unfocused or cloudy thinking
- Changes in your appetite
- Increased alcohol or drug use
- Getting sick more often than usual

8.3 Stress in the Workplace

The root causes of work-related stress often include toxic leadership, long hours, heavy workload, job insecurity, and conflicts with other employees. Symptoms include a drop in work performance, a high degree of sick leave, depression, anxiety, or sleeping difficulties. It is important for employers and leaders to recognize work-related stress as a significant health and safety issue. A company must take reasonable steps to ensure that employees are not subjected to unnecessary stress. Stress signs and symptoms of stress in the workplace are as follows:

- High degree of absenteeism, harming productivity, delay in projects and undermining customer satisfaction.
- Employees can feel sick or dizzy, anxious, worried, or nervous, or become tense, angry, short-tempered, or even despairing.
- Employees may physically react by sweating excessively, experiencing muscle aches, chest pains, blurred eyesight, or itchy skin.

Table 8.1 summarizes the symptoms for stress in six different groups.

8.4 Handling Stress in the Context of Neurodiversity

8.4.1 Effects on Stress

Stress can have a wide range of effects on the body, mind, and behavior. Physically, the negative tension can manifest itself in the form of cardiovascular disorders, migraines, diabetes, heartburn, or stomach ulcers. Psychologically, negative stress can cause feelings of helplessness and exhaustion (Sies, 2019). This can lead to mental illnesses such as depression or anxiety disorders. Negative consequences for dealing with one's social environment and one's job are also possible.

8.4.2 Eustress and Distress

Eustress and distress refer to different types of stress. Eustress feels challenging but manageable and leads to growth, while distress is difficult and has a negative impact (Fig. 8.1).

Eustress is recognized as a positive driver for achieving work goals because it motivates and energizes employees to perform better, fostering a sense of accomplishment and growth. Eustress is usually associated with situations where we feel a sense of control and believe that the stressor is something we can handle (Cooper, 2013). Distress is often linked to situations where we feel a lack of control and see the stressor as a threat. Eustress is increasing the performance, whereas Distress is harming the performance (Helmold, 2022). Eustress is usually associated with situations where we feel a sense of control and believe that the stressor is something we can handle (Sies, 2019). Distress is often linked to situations where we feel a lack of control and see the stressor as a threat.

Table 8.1 Stress symptoms

Emotional symptoms	Physical stress symptoms	Respiratory distress
• Feeling more emotional than usual, especially feeling grumpy, teary, or angry • Feeling anxious, overwhelmed, nervous, or on edge • Feeling sad or depressed • Feeling restless • Trouble keeping track of or remembering things • Trouble getting your work done, solving problems, making decisions, or concentrating	• Headaches • Dizziness • Clenching your jaw and grinding your teeth • Shoulder, neck, or back pain; general body aches, pains, and tense muscles • Chest pain, increased heart rate, heaviness in your chest • Shortness of breath • Feeling more tired than usual (fatigue) • Sleeping more or less than usual • Upset stomach, including diarrhea, constipation, and nausea • Loss of sexual desire and/or ability • Getting sick more easily, such as getting colds and infections often	• Breathing faster than usual • Color changes of your skin, mouth, lips, or fingernails. A blue color around your mouth, lips, or fingernails usually shows you aren't getting enough oxygen. Your skin may also look pale or gray • Grunting when you breath out • A whistling with each breath (wheezing) • Nose flaring • Chest sinking below your neck or under your breastbone with each breath (retractions) • Increased sweating, especially cold, clammy skin on your forehead • Leaning forward while sitting to help take deeper breaths
Cognitive stress symptoms	**Behavioral stress symptoms**	**Chronic stress symptoms**
• Trouble keeping track of or remembering things • Trouble getting your work done, solving problems, making decisions, or concentrating • Feeling less commitment to your work • Lack of motivation • Negative thinking	• Changes in your eating habits; losing or gaining weight • Procrastinating and avoiding responsibilities • Using alcohol, tobacco, or drugs to feel better • Avoiding your friends and family; isolating yourself from others • Failing to meet your deadlines • Increased absences at school or work • Doing your work more slowly • Exercising less often	• Aches and pains • Changes in your sleeping patterns, such as insomnia or sleepiness • Changes in your social behavior, such as avoiding other people • Changes in your emotional response to others • Emotional withdrawal • Low energy, fatigue • Unfocused or cloudy thinking • Changes in your appetite • Increased alcohol or drug use • Getting sick more often than usual

- Eustress increases attention: you are more concentrated and focused on your topic.
- Eustress promotes performance and productivity in negotiations: your muscles are more tense, and your entire body is ready to react. This helps with a tricky task on the computer and also with intensive strength training.
- Eustress motivates: Have you mastered a challenge? The happiness hormone dopamine is released. Congratulations—they increase your motivation.

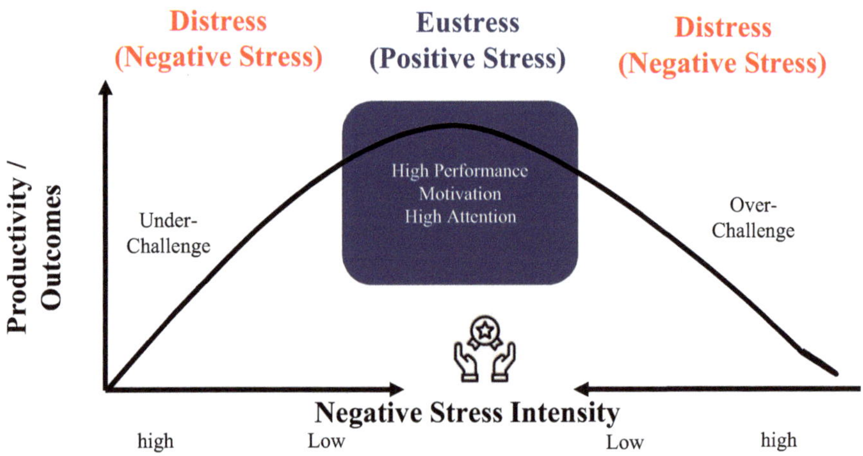

Fig. 8.1 Eustress and distress in relation to outcomes and intensity. Source: Author

- Eustress increases self-confidence: If you solve several problems, this leads to the good feeling of being able to complete tasks successfully.

Figure 8.2 shows the two types of stress, eustress and distress, where eustress is the positive form of stress having a beneficial effect on health, engagement, motivation, performance, and emotional well-being for people and employees, distress leads to anxiety, not-well-being, demotivation, frustration, and hopelessness. Distress is broadly defined as a state of emotional suffering characterized by symptoms of depression (e.g., loss of interest; unhappiness; desperateness) and anxiety (e.g., restlessness; feeling tense). Eustress is perceived as positive and beneficial, while distress is associated with negative outcomes and productivity declines. Eustress may help improve the focus and motivate people to take on new challenges, while distress can lead to anxiety, fear, and other negative emotions.

Eustress is a form of positive stress that motivates and prepares the body for peak performance. When we effectively manage eustress, we build resilience and increase our capacity to handle future stressors. Eustress can help complete the stress cycle by building resilience. By perceiving a stressful situation as a challenge (eustress), individuals can use it to their advantage. When faced with a new stressor, an individual can reconnect to their positive mindset and readjust their coping mechanisms. Distress is a state of severe, prolonged, or overwhelming stress where an individual is unable to cope with or adapt to challenges in their environment. When a stressor (a demand or threat) is perceived, the body's sympathetic nervous system is activated, triggering the release of stress hormones like adrenaline and cortisol. This causes immediate physiological changes such as increased heart rate, blood pressure, and alertness, preparing the body for action. If the stressor persists, the body attempts to adapt and return to a state of normal balance (homeostasis) while remaining on high alert. The parasympathetic nervous system tries to lower the

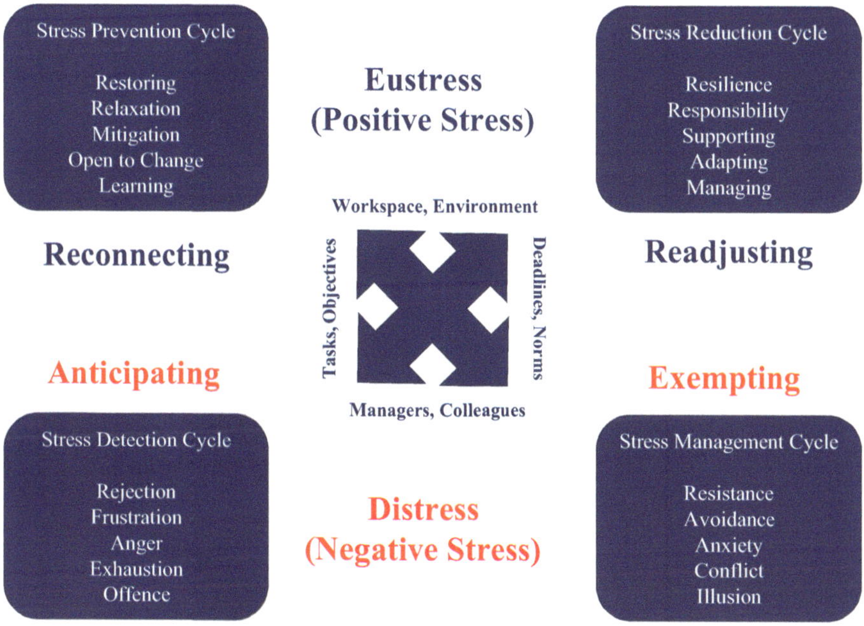

Fig. 8.2 Eustress and distress

heart rate and other bodily functions, but resources are continuously expended to cope with the ongoing demand. This stage occurs if the stress is severe or chronic and the body's resources become depleted. The body can no longer resist the stress effectively, leading to burnout, fatigue, a weakened immune system, and an increased risk of serious health problems like heart disease, anxiety, and depression (Fauquet-Alekhine & Erskine, 2023).

The stress management cycle involves using various techniques and coping strategies to improve the way you react to stressful things in your life and build resilience. Stress management consists of a wide spectrum of techniques and psychotherapies aimed at controlling a person's level of psychological stress, especially chronic stress, generally for the purpose of improving the function of everyday life (Kaluza, 2022). Stress produces numerous physical and mental symptoms, which vary according to each individual's situational factors.

8.5 Recommendations for Effective Stress Management in Neurodivergent Teams: 6A Model

The 6A model depicts in Fig. 8.3 the six elements for effective stress management. The categories include six elements: **anticipate, avoid, alter, adapt, accept, and appreciate**. The 6A model is the extension of the classical A4 model (two elements added are: **anticipate, appreciate**), which has shortcomings.

Fig. 8.3 The 6A model: effective stress management

8.5.1 Anticipate

Anticipatory stress can prospectively and negatively influence diverse outcomes, including cognitive performance and emotional well-being. It has been suggested that perseverative cognitions (e.g., worry, rumination) during the anticipation period constitute a key mechanism driving these effects (Żołnierczyk-Zreda, 2015). Anticipatory stress is experiencing an increased stress response in anticipation of a difficult or unpredictable situation.

It is important to anticipate stress factors and coping ahead. Stress anticipation and management involves thinking about recovery plans and solutions. It is a deliberate and proactive skill, rather than a reactive response that actually impairs your problem-solving abilities.

8.5.2 Avoid

Avoiding stress, stress factors, and stress triggers is important for employees. Taking control of the stress that fills the day by finding ways to make the situation better can help to prevent stress. Creating distance from a stressor is a key stress management technique that can involve physically removing yourself from a situation, mentally re-framing your thoughts, or altering the environment to reduce its impact. This helps to reduce immediate emotional intensity, improve your perspective, and lead to better coping strategies and well-being.

8.5.3 Alter

When stress factors cannot be changed, people can alter the situation. If one cannot avoid a stressful situation, the person can attempt to alter it. It should be figured out what one can do to change things so the problem does not present itself in the future. People should express their feelings instead of bottling them up—communication should raise concerns openly and respectfully. People must be willing to compromise—when someone asks someone else to change the behavior, the person must be willing to do the same. Finally, one should be more assertive and try to deal with problems head-on, doing the best to anticipate and prevent them.

8.5.4 Adapt

Stress adaptation, prophylaxis, and treatment are key strategies in managing the impact of significant stress on the body and mind. The approach often varies depending on whether the goal is to prevent stress-related issues, build resilience, or manage acute symptoms. The human stress response activates the autonomic nervous system and endocrine systems to increase performance during environmental challenges. This response is usually beneficial, improving the chance of overcoming environmental challenges, but costs resources such as energy, power, and time. The human body responds to stressors by activating the nervous system and specific hormones. The hypothalamus signals the adrenal glands to produce more of adrenaline and cortisol and release them into the bloodstream. These hormones speed up heart rate, breathing rate, blood pressure, and metabolism (Rogers & Lucchesi, 2014).

Adaptation toward stress is resilience. Resilience is the ability to adapt to a changing environment, and this adaptive stress response (also known as allostasis) is crucial for survival. Every organism—plants, animals, and humans—has developed a number of mechanisms to adapt to these stressors to ensure growth, survival, and reproduction (Das, 1999). The process of stress adaptation may appear to be simple, but in reality, this is a very complex process and we are only beginning to understand the mechanism of adaptation.

8.5.5 Accept

Accepting certain situations can make uncontrollable stressors more manageable, as it is often our reactions and resistance to external or internal stress factors. By not trying to control or change the uncontrollable, people can let the stressor be and concentrate on attention, time, and energy on things that are more important. From this point, people can choose how they want to act within this reality (rather than trying to change the reality itself).

8.5.6 Appreciate

Cultivating appreciation and gratitude is one of the simpler routes to a greater sense of emotional well-being, higher overall life satisfaction, and a greater sense of happiness. Stress often blinds us to the fact that we deserve to feel proud of ourselves. Appreciating ourselves is not just about celebrating big achievements. Self-worth is more than a victory lap; it is an act of self-love, a gentle acknowledgment of everything we have been through. It helps train the brain to notice and appreciate the little things in life and, in doing so, shifts the life experience tremendously. Gratitude can increase people's happiness and well-being, life satisfaction, even overall health, while decreasing the stuff we all want less of like anxiety, depression, and anger.

8.6 Case Study: Early Warning System for Mental Illness in Porsche

Porsche is a renowned German automobile manufacturer with an emphasis in high-performance luxury sports cars, SUVs, and sedans. The company is actively improving health issues in their workforce. While the associated stigma is slowly being eroded, poor mental health remains one of the biggest issues affecting society today. Around 20 percent of the working population will go through a phase at least temporarily in their lives in which they reach their psychological limits—or go beyond them. And the trend is increasing. As is so often the case in medicine, the causes are not one-dimensional. Many factors play a role in mental illnesses, such as genetics, resilience, family and work-related stress, or social environment. Often, a personal crisis is the last straw. Around eight percent of disabilities are due to mental illness. Paying attention to the health of the workforce is important, especially in times of demographic change and a rapidly changing working world. That is why Porsche has now developed an early warning system for mental illness (Pankow, 2019). It is intended to prevent possible chronic diseases caused by excessive psychosocial stress.

The so-called Allostatic Load Index uses just five variables to determine a key figure for stress-related health problems. The Association of German Company and Factory Doctors has now awarded the Innovation Prize for this so-called practical index. Allostatic load refers to the overuse and wear and tear effects that occur in an organism due to repeated or long-term stress. Allostatic load was developed in the USA in 1993. The index is based on just five values:

- Diastolic blood pressure: This is the lower measurement value after the heart muscle relaxes.
- Abdominal circumference.
- LDL cholesterol.
- Long-term blood sugar.
- Heart rate variability.

It can therefore be calculated from any normal check-up by your family doctor or company doctor and is therefore comparatively inexpensive. The variables are also based on threshold values. This means that we have defined a value for each variable, and if this value is exceeded, an index point is included in the calculation (Porsche, 2024).

References

Adamsen, B. (2023). *Work identity. A neurophilosophical perspective*. Springer.

Cooper, C. L. (2013). From stress to wellbeing. In *The theory and research on occupational stress and wellbeing* (Vol. 1). Palgrave McMillan.

Das, D. K. (1999). *Stress adaptation, prophylaxis and treatment*. Springer.

Fauquet-Alekhine, P. ans Erskine, J. (2023). The Palgrave handbook of occupational stress. Springer .

Helmold, M. (2022). Leadership. Agile, virtuelle und globale Führungskonzepte in Zeiten von neuen Megatrends. Springer .

Kaluza, G. (2022). *Calm and confident under stress. The stress competence book: Recognize, understand, Manage Stress*. Springer.

Pankow, G. (2019). Porsche-Mediziner erhält Innovationspreis. Psychische Erkrankungen: Technik für die deutsche Wirtschaft. Porsche entwickelt Frühwarnsystem. Produktion. https://www.produktion.de/wirtschaft/gehalt-karriere/psychische-erkrankungen-porsche-entwickelt-fruehwarnsystem-126.html.

Porsche. (2024). Frühwarnsystem ermittelt Risiko für psychische und chronische Erkrankungen. Retrieved December 30, 2024. https://newsroom.porsche.com/de_CH/2019/unternehmen/porsche-fruehwarnsystem-risiko-psychische-chronische-erkrankungen-daniel-mauss-leiter-gesundheitsmanagement-19441.html.

Rogers, L. K., & Lucchesi, P. A. (2014). Stress adaptation and the resilience of youth: Fact or fiction? National Library of Medicine. Retrieved December 30, 2024. https://pmc.ncbi.nlm.nih.gov/articles/PMC5243201/.

Sies, H. (2019). Oxidative stress. Eustress and Distress. Academic Press. .

Żołnierczyk-Zreda, D. (2015). The effects of worksite stress management intervention on changes in coping styles. *International Journal of Occupational Safety and Ergonomics*, 465–482. Retrieved December 28, 2024. Published online: 8 Jan 2015. https://www.tandfonline.com/doi/abs/10.1080/10803548.2002.11076548

Motivation and Incentives

9

9.1 Motivation: Trigger for Action and Behavior

Motivation refers to a process of inducing and stimulating an individual to act in certain manner. In the context of an organization, motivation implies encouraging and urging the employees to perform to the best of their capabilities so as to achieve the desired goals of the organization. Motivation is not necessarily linked to job satisfaction, but to a match between incentives (from management) and motives (of the individual employee) (Helmold, 2021). For an employee to be motivated, four aspects must be in harmony as shown in Fig. 9.1: individual triggers, personal capabilities, social boundaries, and situational enablement (Helmold, 2022). Whereas the individual triggers (want to do) and personal capabilities (able to do) are internal triggers, social boundaries (allowed to do) and the situational enablement (possible to do) are external triggers (Frederick, 1968).

Individual Triggers (want to do): Individual triggers refer to the driver and values that drive a person to engage in activities or tasks, influenced by factors such as job design, autonomy, variety, and meaning, which can impact their well-being, job satisfaction, and productivity. Individual triggers are desires and motives that lead to the action and engagement.

Social Boundaries (allowed to do): Social boundaries are norms, rules, and the environment, which trigger actions and behavior. Social boundaries define the freedom and the limits, in which people can act. Boundaries help individuals and groups to take care of and prioritize the needs and the well-being. Boundaries help people strive for innovation and action, and maintain a work–life balance that is healthy for individuals and for others around them.

Situational Enablement (possible to do): Situational enablement helps individuals with the freedom and independency to do certain tasks or jobs. The situational enablement contains attitudes, conditions, and support in society and organizations.

M. Helmold, M. Martensen, *Leadership in Neurodiversity and Neurodivergence*, Diversity and Inclusion Research, https://doi.org/10.1007/978-3-032-09343-1_9

Fig. 9.1 Triggers for motivation

Personal Capabilities (able to do): Capabilities refer to an individual's psychological and physical ability to participate in an activity. Personal capabilities include skills and attitudes.

9.2 Willingness/Capability Matrix (Skill/Will Matrix)

The willingness/capability matrix is a suitable leadership tool for non-neurodiversity and neurodiverse teams. It is a simple but powerful 2 × 2 performance management and employee engagement framework to help managers identify and adopt the most appropriate leadership approach for each team member and situation, based on two metrics: their skill and will (task and physiological maturity) levels. Skill relates to the individual's capability and experience (core capability), while will relates to motivation (willingness, enthusiasm, and self-drive) (Hersey & Blanchard, 1992).

Low Skill, High Will: Individuals in the top-left corner are highly motivated but lack the necessary skills or experience. They are eager to learn and contribute, but they need guidance and training to develop their abilities.

Low Skill, Low Will: Individuals in the bottom-left quadrant lack both skills and motivation. They may be new to the role or organization, or they may be disengaged for some other reason. These people need the most help.

High Skill, Low Will: In the bottom-right quadrant, individuals have high competence, but lack the motivation to put their skills to use. They know how to do the job, but for one reason or another, they're not committed. This could be due to a

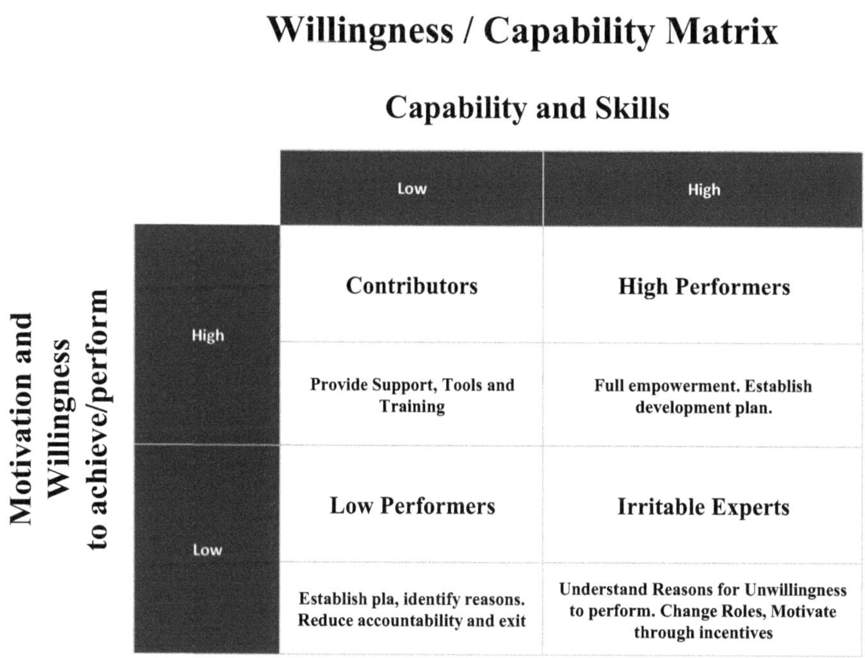

Fig. 9.2 Willingness/capability matrix

variety of factors, such as lack of confidence or dissatisfaction with the job. These individuals often need support in finding their motivation, or addressing obstacles that are blocking their commitment.

High Skill, Low Will: The heady top-right corner is the quadrant of self-reliance. Individuals in this quadrant are experienced, skilled, and motivated. They have a clear understanding of what needs to be done and the drive to do it. They are often the "stars" of the team, requiring little supervision and providing significant contributions.

The matrix was derived from the Situational Leadership Model developed by Paul Hersey and Ken Blanchard in 1969, which is based on the study of relationships between leaders and followers and provides a useful framework to analyse leadership situations based on the performance readiness level that a follower exhibits in performing a specific task, function, or objective (Kenton, 2024) (Fig. 9.2).

9.3 Two-Factor Theory by Herzberg

9.3.1 Scope and Definition

The two-factor theory (also called motivator-hygiene theory) by Frederick Herzberg et al. (1959) is a content theory of motivation, especially work motivation. Together with Maslow's hierarchy of needs, it is one of the best-known motivation theories, whose common feature is that they offer a classification of the motives of human action.

As the name suggests, Herzberg distinguishes between exactly two types of influencing factors. Firstly, factors that are related to the content of the work (motivators) and secondly, factors that are related to the context of the work (hygiene factors). Content factors include, for example, having responsibility or gaining recognition; context factors can be pay and external working conditions.

However, satisfaction and dissatisfaction do not represent the two extreme manifestations of a characteristic, but should be viewed as two independent characteristics. The "hygiene factors" (dissatisfied–not dissatisfied) and the "motivators" (not satisfied–satisfied) represent these two areas. According to the theory, both characteristics must be present in order to experience job satisfaction.

Satisfaction does not necessarily exist if there are no reasons for dissatisfaction.

9.3.2 Hygiene Factors

Herzberg defines hygiene factors as those factors which, when positive, prevent the emergence of dissatisfaction but do not contribute to or create satisfaction. These factors are often not noticed or taken for granted. But if they are not present, this is perceived as a deficiency. Hygiene factors include in particular the following:

- Remuneration and salary
- Personnel policy, leadership style
- Working conditions including autonomy and support
- Interpersonal relationships with employees and managers
- Job security, assurance
- Influence on private life

The concept of the hygiene factor reveals the essential idea of the theory. The factors of dissatisfaction should be removed from the human environment, just as in medical hygiene health risks are removed from the human environment in order to prevent illness. As an example, dissatisfaction arises when cooperation with others does not work or the company structure seems "impossible." If all of these extrinsic aspects are sufficiently well developed, however, satisfaction does not arise, but rather a neutral state of experience, which is referred to as non-dissatisfaction (state: disinterest or indifference). Favorable hygiene factors do not make people happy, they "just" do not make them unhappy. This approach also explains why wages that are too low make people unhappy, but why the motivation and satisfaction of employees cannot be increased indefinitely through salary.

9.3.3 Motivators

According to Herzberg, motivators influence the motivation to perform and primarily come from the content of the work. Motivators therefore change satisfaction, but their absence does not necessarily lead to dissatisfaction. The focus here is on striving for growth and self-satisfaction. Motivators include in particular the following:

- Work performance and success
- Recognition and reward
- Work content
- Responsibilities, roles
- Advancement and promotion
- Personal growth and opportunities

The combination of hygiene factors and motivators creates four possible situations:

- High hygiene and high motivation: The ideal situation in which employees are highly motivated and have few complaints.
- High hygiene and low or no motivation: The employees have hardly any complaints, but are poorly motivated (mercenary mentality).
- Low or no hygiene and high motivation: The employees are motivated, but have many complaints. The job is exciting and challenging, but the working conditions are not so good.
- Low or no hygiene and low or no motivation: The worst situation. Unmotivated employees with many complaints.

Some of the motivators can also act as hygiene factors, i.e. become a matter of course. Conversely, hygiene factors can gain in importance and become motivators if they have been absent for a long time. The classification of individual factors in the group of hygiene factors or motivators therefore depends in part on the specific situation and the experience background of the individual and society as a whole (Fig. 9.3).

9.4 Maslow's Hierarchy of Needs

Maslow's hierarchy of needs is a model that organizes human needs in a hierarchical order, starting with physiological needs (hunger, thirst, etc.), followed by safety needs, social needs, ego needs, and, finally, the need for self-actualization. According to Maslow, a higher ranking need only becomes relevant when the needs below it are satisfied.

The physiological needs (food, warmth, etc.) are the most basic and powerful of all: The needs that are usually used as the starting point of motivation theory are the so-called physiological drives.

Fig. 9.3 Two-factor theory by Herzberg

Then, provided that the physiological needs are largely satisfied, security needs follow. Security needs (2nd stage) are understood to mean security; stability; safety; protection; freedom from fear; need for structure, order, law, boundaries; protective power. Next, social needs arise (3rd stage).

When both the physiological and security needs are satisfied, the needs for love, affection, and belonging will emerge.

As the process continues, needs for respect (4th stage) and self-actualization (5th stage) can be satisfied. Maslow claims that there are real psychological and functional differences between the "higher" and "lower" needs.

Although the higher needs are specific to humans (in contrast to, for example, animals), they are not absolutely necessary for their survival. The needs can also be differentiated according to deficit needs (essential needs, 1st–4th level) and growth needs (higher needs); deficit needs must be met in order for satisfaction to arise; the additional fulfillment of growth needs means happiness that goes beyond satisfaction. Only when the deficit needs are definitely satisfied and satisfaction on a physical level in the form of longer life expectancy, less illness, and a better nutritional situation occurs, do the growth needs, which were previously subjectively less urgent, come to the fore. Their satisfaction in turn leads to deeper happiness, serenity, richness of inner life, and increased individuality. Furthermore, the higher needs and their satisfaction have desired civic and social consequences. Scientists criticize the lack of empirical confirmation of the theory, its culture-specific focus on the American middle class, the imprecise delimitation of need categories, and the vague definition of self-actualization. In addition, the assumption that human behavior is always controlled by only one need is viewed as simplistic (Helmold, 2022) (Fig. 9.4).

9.5 Incentives in the Context of Neurodiversity

Studies show that organizations that hired neurodivergent workers attained improved overall employee morale and positive impacts on organizational culture. Many neurotypical workers appreciate learning about neurodiversity and experience pride in working at inclusive workplaces that embrace diverse ways to think, learn, and

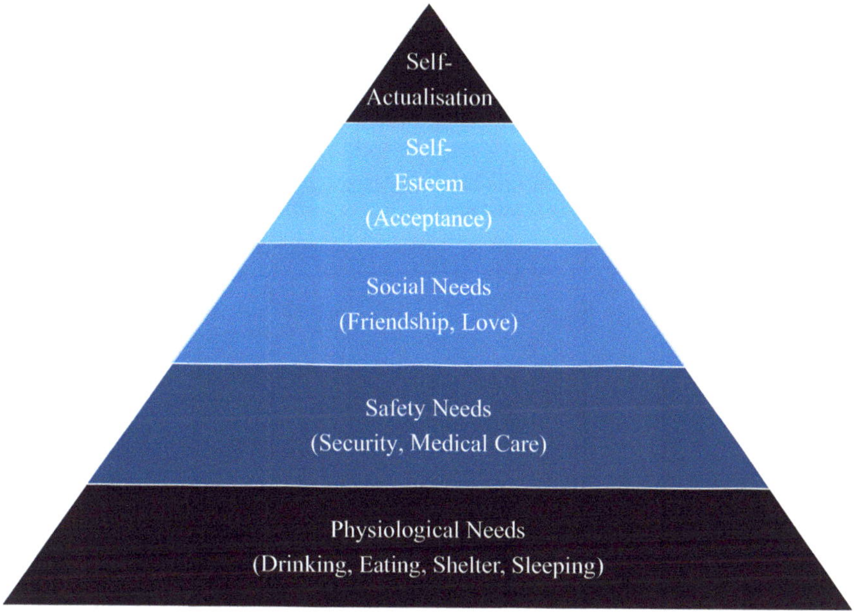

Fig. 9.4 Maslow's hierarchy of needs. Source: Author

perceive the world. This type of workplace culture helps them better understand and connect with their neurodivergent colleagues, as well as other coworkers, as collaborators and friends. Many best practices for working with people with disabilities in general can apply equally to working with neurodivergent people as well. Likewise, neurodivergent workers can greatly benefit from their full inclusion in a diverse workplace. Many neurodivergent people, including autistic people, share common experiences of discrimination and stigma in work settings. Efforts to support greater inclusion in the workplace can help address this stigma, especially when neurodivergent workers are actively included in efforts to improve workplace inclusion.

In addition, neurodivergent workers report greater confidence and self-esteem when their job duties align well with their set of talents, strengths and skills—much like other workers. Workplace inclusion may also help many neurodivergent people gain confidence in other areas, such as participating in their community activities and interpersonal relationships. Thus, providing needed access to workplace accommodations and supports can help drive this comfort and support retention of neurodivergent workers (Earn, 2024). The biggest goal for a neurodiverse person is feeling empowered by their differences, not lesser for having them. Figure 9.5 depicts monetary and non-monetary incentives. As a concept of strategic importance to the HR department and leadership, it must be expanded to include and accommodate neurodiverse talent in the workforce. Neurodivergent people should be paid equally; however, there are more opportunities for HR and leaders to embrace neurodiversity in enterprises. These can be medical or financial support

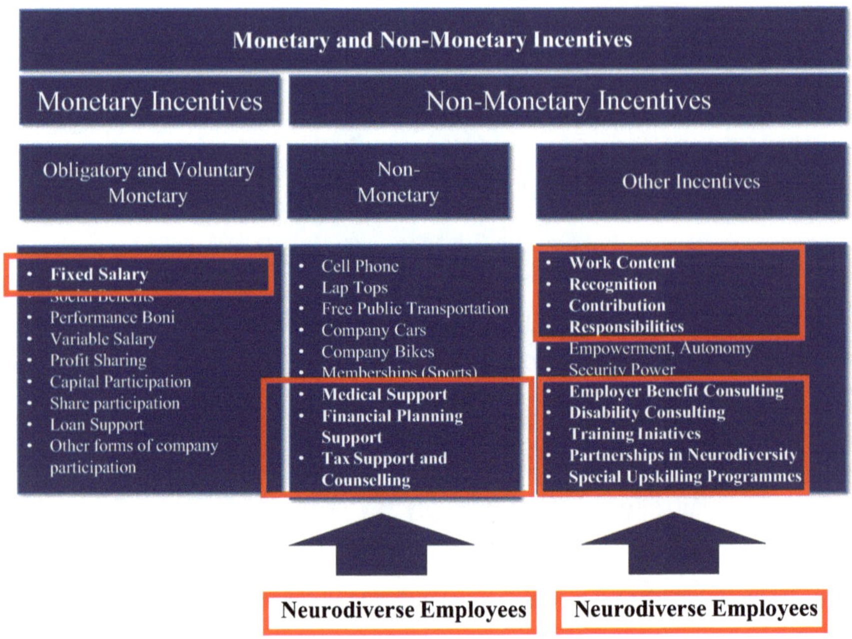

Fig. 9.5 Incentives in the context of neurodivergent people

planning combined with employer benefit consulting through experts, associations, or consultants (Howlett, 2022) (see Fig. 9.5). Flexible work arrangements, such as remote work, floating start times, and compressed workweeks, can significantly benefit neurodivergent employees. This flexibility allows them to engage in upskilling activities at times and in environments that best suit their learning styles (Palumbo, 2024).

9.6 Case Study: Inclusion in Google, Talent Engagement

Googlers with disabilities make a critical contribution to the entire company. With support from the disability community at Google, the company has found new ways to engage and hire more people with disabilities. Google is working tirelessly to make Google a place where people with disabilities can reach their full potential. The Talent Engagement team in Europe, the Middle East, and Africa hosted its first Disability Conference for nearly 2000 students, graduates, and industry professionals with disabilities. The event created a place where people with disabilities and their allies could interact with each other and with Google. To increase the representation of people with disabilities, Google launched gReach in Asia. gReach is a paid upskilling program for students and industry professionals with disabilities, providing all participants with hands-on training and work experience within Google teams.

In partnership with Google's Disability Alliance Employee Resource Group, Google launched a new global e-learning program to help hiring managers better assess the potential of candidates with disabilities and neurodiversity. It is now available to all Google employees. The company also works continuously with the Disability Alliance to create inclusive workplaces around the world that address the challenges of Googlers with disabilities. In a world of hybrid work models, Google recognizes that our policies and strategies cannot be one-size-fits-all. In 2022, we established new global policies to create spaces that are accessible and inclusive for all employees, without exception.

References

Earn. (2024). Employer assistance and resource network disability on inclusion. Employee benefits of neurodiversity. Learn how hiring neurodivergent workers can benefit all employees. Retrieved December 30, 2024. https://askearn.org/.

Frederick, H. (1968). One more time: how do you motivate employees? *Harvard Business Review, 46*(1), 53–62.

Frederick, H., Bernard, M., & Snyderman, B. B. (1959). *The motivation to work. 2. Auflage.* Wiley.

Helmold, M. (2021). New work, transformational and virtual leadership. Lessons from COVID-19 and other crises. Springer.

Helmold, M. (2022). *Leadership. Agile, virtuelle und globale Führungskonzepte in Zeiten von neuen Megatrends.* Springer.

Hersey, P., & Blanchard, K. H. (1992). *Management of organizational behavior. Utilizing human resources.* Prentice Hall.

Howlett, E. (2022). The evolution of employee benefits to better support neurodiversity. Rewards and Employee Benefits Association. Retrieved December 30, 2024. https://reba.global/resource/the-evolution-of-employee-benefits-to-better-support-neurodiversity.html.

Kenton, W. (2024). The situational leadership model: How it works. Retrieved December 30, 2024. https://www.investopedia.com/terms/h/hersey-and-blanchard-model.asp#:~:text=The%20Hersey%2DBlanchard%20Model%20suggests,they%20lead%20and%20their%20abilities.&text=The%20Hersey%2DBlanchard%20Model%20is,Situational%20Leadership%20Model%20or%20Theory.

Palumbo, J. J. (2024). How companies can Foster the strengths of neurodivergent employees. Forbes. Retrieved December 30, 2024. https://www.forbes.com/sites/jenniferpalumbo/2024/06/27/how-companies-can-foster-the-strengths-of-neurodivergent-employees/.

Emotional Intelligence (EI), Active Listening (AL), and Neuro-Linguistic Programming in Neurodiverse Teams

10

10.1 Emotional Intelligence (EI)

Emotional Intelligence (EI) is the ability of leaders and managers to manage both their own emotions and understand the emotions of the work team employees. For neurodiverse individuals, EI can manifest in unique ways, fostering communication styles that are transparent and direct, often leading to clarity in tasks and deliverables. EI is the ability to correctly assess feelings (in oneself and others) and human relationships and to act accordingly.

In recent years, the emphasis on fostering emotional intelligence (EI) within neurodiverse work environments has gained significant traction. A study by the Harvard Business Review found that companies with high emotional intelligence capabilities retain 80% of their employees, compared to just 40% in less emotionally intelligent firms. Imagine a workplace where diverse minds, including those on the autism spectrum, are not only welcomed but supported in expressing their emotions constructively. By implementing training programs focused on EI, organizations can create a culture of understanding and empathy, leading to increased collaboration and a 32% boost in overall productivity, as reported by the American Psychological Association (Vorecol, 2024).

The term "emotional intelligence" became popular through the book of the same name by the American journalist and psychologist David Goleman from 1995. Goleman sees emotional intelligence as an important prerequisite for a professional career and successful leadership. As early as 1920, the famous psychologist and intelligence researcher Edward Thorndike used the term "social intelligence" for the ability to correctly assess and guide other people. Goleman later distinguished four areas of competence in emotional intelligence:

- Emotional intelligence as a learnable ability (Ability EI), e.g. the ability to correctly perceive feelings

M. Helmold, M. Martensen, *Leadership in Neurodiversity and Neurodivergence*, Diversity and Inclusion Research, https://doi.org/10.1007/978-3-032-09343-1_10

- Emotional intelligence as a basic characteristic (Trait EI)
- Emotional intelligence as a mixture of motivation, skills, and personality traits (Mixed EI, 10.2 Emotional Intelligence)
- Social awareness: Perceiving and understanding the feelings and relationships of others
- Relationship management: Understanding and influencing interpersonal relationships
- Current research distinguishes among three models of emotional intelligence:
- Emotional intelligence as a learnable ability (Ability EI), e.g. the ability to correctly perceive feelings
- Emotional intelligence as a basic characteristic (Trait EI)
- Emotional intelligence as a mixture of motivation, skills, and personality traits (Mixed EI, 10.2 Emotional Intelligence)

10.2 Advantages of EI

The special thing about emotional intelligence is that it is about dealing with yourself as well as with other people. Emotional intelligence describes self-management and self-awareness, on the one hand, and skills and abilities in dealing with other people, on the other. The following skills are particularly important for emotional intelligence in neurodiverse teams (Fig. 10.1):

- Self-awareness: This means the realistic assessment of one's own personality, that is, recognizing and understanding one's own feelings, needs, motives, and goals, but also awareness of personal strengths and weaknesses. It is about knowing yourself well in order to be able to assess how you react in certain situations, what you need and where you still need to work on yourself.

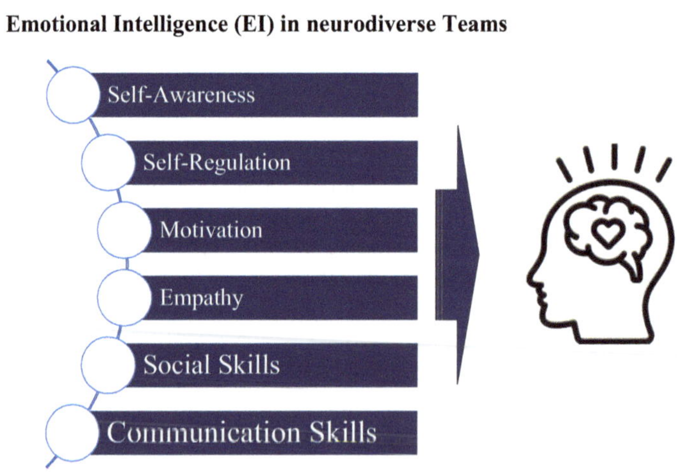

Fig. 10.1 EI in neurodiverse teams. Source: Author

- Self-control: Self-control is the ability to influence and control your own feelings and moods through an internal dialogue. With this ability, we are no longer simply at the mercy of our feelings, but can influence them constructively. An example: If something makes us angry, we can calm ourselves down through our inner dialogue and can then react much more appropriately than if we are not able to control ourselves.
- Motivation: Being able to motivate yourself means being able to continually develop a willingness to perform and enthusiasm from within yourself. This ability is particularly helpful in phases when a project becomes difficult or when things don't go as planned. Those who can motivate themselves always find the strength to carry on and also have a higher frustration tolerance, i.e. the ability to endure frustration and still carry on.
- Empathy: Empathy means successfully dealing with others. This means the ability to put yourself in the shoes of other people and to react appropriately. It is about perceiving and accepting other people as they are. Acceptance does not automatically mean approval. Accepting other people means treating them with respect and understanding their actions and thoughts.
- Social competence: Social competence is understood to mean, for example, the ability to establish contacts and relationships with other people and to be able to maintain such relationships over the long term. This means good relationship and conflict management, but also leadership qualities or the ability to form and lead functioning teams.
- Communication skills: Good communication skills are essential for emotional intelligence. This means two things: on the one hand, the ability to express oneself clearly and understandably and thus to communicate one's concerns clearly and transparently; on the other hand, it means the ability to listen actively and attentively to other people and to understand and classify what they say.

Goleman's EI concept has been criticized many times, among other things because it mixes learnable skills with basic personality traits. Goleman has also been accused of simply selling existing older concepts under a new name. Some researchers deny that emotional intelligence is even an independent concept. They believe that emotional intelligence largely results from the expression of the five basic personality dimensions (Big Five). Despite this criticism, the concept has become established in the private sector. Many companies now select applicants and managers based on EI characteristics such as empathy and self-management. Numerous studies prove them right, showing time and again that it is not ability EI that is decisive for professional success, but rather mixed EI in the sense of Goleman. Emotional intelligence enables managers to create a better working atmosphere, for example, and thus increase productivity by up to 20% (Goleman, 2008).

10.3 Active Listening (AL)

10.3.1 Scope and Definition

Successful leaders carefully check whether the negotiation opponent is satisfied in his positions, too. Taking the self-esteem of the employee is a key factor in successful negotiations. The active and analytical listening method is a suitable concept to understand the other negotiation motives and interests. Active and analytical listening in negotiations is understood to be the emotional (affective) reaction of a conversation partner to the message of the negotiating partner. Active listening is the emotional and factual reaction of the listener to the various aspects of the speaker and what is said. In essence, it is about not just listening passively and letting the other person speak, but also remaining active, even when you are not speaking yourself. The information received is also analyzed through questions, questions for comprehension, and the reaction of the negotiating partner. Information can be verified by asking similar counter-questions that probe the source's credibility, such as How do they know? or Are they biased?. Other methods include seeking out different perspectives, comparing information across multiple credible sources, and distinguishing between facts and opinions. There are also suitable strategies, tactics, and questioning techniques. Active and analytical listening means an open, active, respectful, and empathetic attitude toward the conversation partner and the content of the conversation. This means that you listen carefully to your conversation partner, agree with them and, if necessary, ask questions about what they are saying, but without interrupting the other person. Active listening is one of the most effective communication techniques for successful negotiations. Active listening is an essential part of modern and innovative communication in teams. This method aims to obtain as much information as possible from the employees and to reveal as little information about yourself. On the other hand, the concept also serves a deeply human need, namely to be understood. Anyone who can convey this feeling to the team members has as good as won.

10.3.2 Observation and Perception

In the first step, the behavior and statements of the negotiating opponent are observed and perceived. It is important not only to concentrate on the factual statements of the conversation partner, but also to observe and interpret their posture, facial expressions, and gestures (Helmold, 2022). Eye contact and pauses help the conversation partner to organize their thoughts and to articulate themselves in detail in a relaxed atmosphere.

10.3.3 React

The second step is the reaction. Agreement through a slight nod is a way of giving the negotiating opponent the feeling that you are listening attentively. In this step, the method of mirroring is usually helpful. In the conversation, you can mirror body language and content. Light and short paraphrasing can also be a suitable technique.

10.3.4 Understanding and Interpreting

In the third step, the statements of your conversation partner are interpreted and interpreted against the background of your own values and experiences. Here, the factual and emotional parts of the speech of the negotiating opponent are accepted, interpreted, and interpreted based on one's own and social values.

10.3.5 Answers

The last step is to react with a careful and well-chosen answer. The reaction to the speech is to reproduce what was said in an appropriate form. In doing so, one focuses on the points that, from one's own perspective, are important and useful for the further course of the conversation. When "paraphrasing," one repeats information or arguments from the communication partner in one's own words. It is best to formulate the question form here in order to enable possible correction. ("Did I understand you correctly that.....? Did you mean....? Did you mean this....?")

Here, one addresses the emotional, hidden messages of the conversation. One expresses the presumed moods and emotions of the conversation partner in one's own words. This only applies if the conversation situation allows it. You should make sure that the nuances are set correctly. It is not what is said that is interesting, but above all, how it is said. This refers to the tone of voice, the way of speaking, and the speed of speech. It is important that this is your perception and that you are prepared to accept corrections and objections.

Practical tip: Combine active listening with a pinch of humor. Humor makes your life easier, it humanizes your communication and creates closeness. People who act and react with humor are seen as more likeable and trustworthy. Communication can be fun! Humor is also a good way to let off steam, especially when things aren't going so well (Fig. 10.2).

10.3.6 Advantages of AL

When actively listening, the listener reflects both factual and emotional aspects of his or her speech back to the other person. Especially when dealing with difficult and emotional topics, this form of conversation builds mutual trust and thus improves

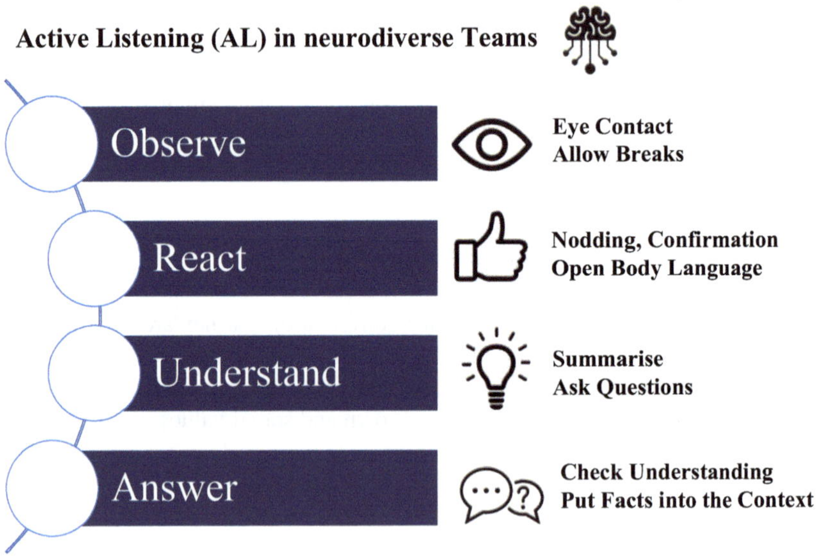

Fig. 10.2 Active listening. Source: Author

communication. Active listening also makes it possible to initiate a change in the other person's behavior.

Communication: If you let your opponent talk more, you will not only receive gratitude, but also many tips on how to conduct the negotiation in the final phase. By then, you have already achieved three important things with active listening: your own competence is recognized, you are valued as a person, and the trust of your opponent has increased. This results in a number of advantages in further communication.

Influencing orientation: Analytical listening has grasped the opponent's main points. Now it is just a matter of linking it to your own objectives and weighting it in your own interests.

Moderating the content: The interim summaries have put the negotiating partner, who has been actively listening, into a second role: that of the moderator. This naturally also includes a neutral position, which the negotiating partner now subliminally acknowledges. The moderator, however, takes over the leadership of the conversation.

Controlling the results: If 90% of the negotiation has been conducted using an integrative strategy, this will not change in the last few minutes. The negotiating partner has been drawn into the active listener's pull and will only make minor attempts to change the emerging negotiation result. If the active listener has done good preparatory work, he can even show conciliation here. The negotiating partner leaves the negotiation with a good feeling, even though he has not achieved his goals. But the most important thing is: he will always seek negotiations with the active listener. This interaction is fundamental to successful negotiation and should be part of every negotiation training!

10.3.7 Recommendations for AL in Neurodiverse Teams

Show acceptance and agreement: As everyone knows, communication only works as long as one party is speaking and the other is listening. But is it enough to simply listen? Especially with more complex topics? Everyone has caught themselves drifting off while the other person is talking to them. How do you prevent this and how do you get the most information out of a conversation?

Nobody likes talking to a wall. So, actively convey to your conversation partner that you are interested in listening to them. Techniques and methods include maintaining eye contact, nodding, repeating what has been said, asking questions.

Remain open and empathetic toward your employee and conversation partner: Create an open and relaxed atmosphere. Show that you are open to other opinions and do not judge in a biased way. Your counterpart should have the feeling that you have compassion for them and that you understand other ways of thinking. Avoid interrupting, arguing, or correcting facts. To avoid interrupting, listen actively, and only speak when the other person is finished, even if you have a point to make. Techniques include using a physical cue like a finger to note your thought, jotting down ideas to avoid forgetting them, or taking a moment to breathe and match the speaker's pace. Active listening techniques include giving undivided attention, using positive non-verbal cues like eye contact and nodding, paraphrasing and summarizing to confirm understanding, and asking clarifying questions. It is also important to avoid distractions and interrupting the speaker.

Avoid distractions: AL is the exact opposite of inactive listening. You may be listening, but your mind wanders and the information literally goes "in one ear and out the other." Active listening techniques create a quiet space for talking by eliminating distractions, encouraging full focus on the speaker, and using verbal and non-verbal cues to show engagement without interrupting. Key methods include finding a quiet place, putting away your phone, making eye contact, maintaining open body language, and resisting the urge to formulate your own response while the other person is speaking.

Don't allow yourself to be under time pressure: As soon as you are under time pressure, you stop listening. While you are listening, you think about your answer in order to end the conversation as quickly as possible. In doing so, you lose half of the information. Take the time for a longer conversation, especially in today's times when digital communication is increasingly taking over.

10.4 Neuro-Linguistic Programming (NLP) as a Leadership Tool in Neurodiverse Teams

10.4.1 Advantages of NLP in the Context of Neurodiversity

Neuro-Linguistic Programming (NLP) sees itself as a toolbox for communication and leadership. NLP strategies can play a pivotal role in facilitating communication and understanding within neurodiverse teams, helping to bridge gaps and promote communication.

NLP arose from the approach of analyzing the behavior of particularly successful people in order to derive action measures and leadership approaches (Holzfuss & Frank-Holzfuss, 2020). Neuro-linguistic programming describes the essential processes of how people perceive themselves and their environment, process this information, feel and act on this basis, and communicate, learn, and change with each other accordingly. NLP can be used in many ways in life, e.g., to change undesirable behavior, improve one's own verbal and nonverbal communication, create understanding and being understood, as support in educational matters and to intensify professional and private relationships. NLP offers very effective methods and techniques to individually assess your own leadership style in the area of tension between humanity and understanding the interests of the employee, as well as the needs and economic interests of the company, and to make the best possible decisions in the interests of the ideal overall result (Buchner, 1994). NLP not only provides managers with strategies for dealing with individual employees, but also techniques for team building and personal development. Through individual coaching, managers learn to reflect on their own behavior during their daily work and to implement appropriate behavior in line with the respective goals. Neuro-linguistic programming (NLP) for managers includes the following:

- Assessing people correctly (body language, facial expressions, eyes)
- Questioning facts
- Understanding problems quickly
- Conflict resolution strategies
- Recognizing people's needs
- Understanding different worldviews
- Understanding roles and using them correctly
- Understanding situations with 1-2-3 perception positions
- Quick-wittedness techniques
- State management (dealing with myself)
- Understanding projects as a system (Fig. 10.3)

10.4.2 Examples of NLP in the Context of Neurodiversity

From the perspective of educators and therapists, NLP tools can be tailored to support language development and social skills training. For instance:

1. Customized Learning Tools: NLP-powered applications can create personalized learning environments that adapt to the individual's language abilities and learning style. For example, an App might use NLP to modify the complexity of language in real time, matching the user's comprehension level.
2. Social Interaction Enhancements: Interactive NLP systems can simulate social scenarios, providing a safe space for individuals with autism to practice and develop social skills. An example is a virtual assistant that uses NLP to engage in conversation, teaching nuances such as sarcasm or idioms.

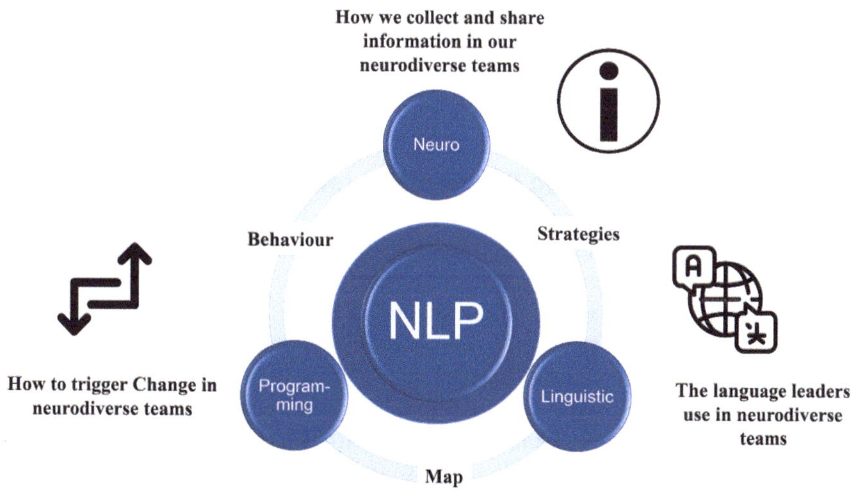

Fig. 10.3 NLP in neurodiverse teams. Source: Author

3. Emotion Recognition: NLP combined with emotion recognition technologies can help individuals with autism interpret and respond to emotional cues in text, aiding in emotional literacy and empathy development.

From the workplace perspective, NLP can facilitate the integration of neurodiverse individuals into business environments:

1. Communication Aids: NLP tools can assist in translating complex instructions into simpler, more manageable tasks, making workplace communication more accessible for employees with autism.
2. Productivity Enhancements: By automating routine tasks with NLP, employees with autism can focus on areas where they excel, often related to pattern recognition and attention to detail.
3. Inclusive Design: NLP can be instrumental in designing inclusive workspaces that cater to diverse communication needs, promoting a culture of acceptance and understanding.

Individual employees with neurodiverse disorders can find empowerment and independence through NLP technologies:

1. Self-Advocacy: NLP tools can help individuals express their needs and preferences more clearly, fostering self-advocacy and independence.
2. Community Building: Online platforms powered by NLP can connect individuals with autism, creating communities where they can share experiences and support each other.

3. Personalized Interfaces: NLP can customize user interfaces to match the cognitive and sensory preferences of individuals with autism, making technology more accessible and enjoyable.

10.5 Case Study: Using Emotional Intelligence in SAP

One leading tech firm, SAP, has successfully embraced this approach by integrating EI training within their neurodiversity hiring initiatives. By teaching managers how to recognize and appreciate the unique emotional expressions of their neurodiverse teammates, SAP reported a 22% increase in team cohesion, allowing for innovative problem-solving (Vorecol, 2024). According to a 2020 study from the Neurodiversity Hub, companies that actively cultivate emotional intelligence in their neurodiverse workforce experience a remarkable 30% lower turnover rate. These statistics showcase that fostering emotional intelligence is not just a noble pursuit; it is a strategic business decision that enhances employee well-being and drives overall organizational success.

References

Buchner, D. (1994). *NLP im Business: Konzepte für Veränderungscoaching*. Springer.

Goleman, D. (2008). *Soziale Intelligenz*. Knaur München.

Helmold, M. (2022). Leadership. Agile, virtuelle und globale Führungskonzepte in Zeiten von neuen Megatrends. Springer .

Holzfuss, B., & Frank-Hollzfuss, J. (2020). Das NLP Practitioner Prüfungswissen kompakt: Die Wissensessenz aus zehn Jahren erfolgreicher NLP Ausbildung. Unabhängig publiziert.

Vorecol. (2024). Exploring the link between neurodiversity and emotional intelligence development through 360 degree feedback. Retrieved December 30, 2024. https://vorecol.com/blogs/blog-exploring-the-link-between-neurodiversity-and-emotional-intelligence-development-through-360degree-feedback-194743#:~:text=For%20neurodiverse%20individuals%2C%20emotional%20intelligence,clarity%20in%20tasks%20and%20deliverables.

Neurodiversity: Implications for Leadership, Training, and Development

11

11.1 The Significance of Leadership in Personnel Development

Leadership plays a pivotal role in driving organizational success, particularly in the domain of personnel development. While management ensures operational efficiency and stability, leadership fosters growth, innovation, and alignment with strategic goals. Kotter (1990) eloquently differentiates management from leadership, emphasizing their complementary functions: management focuses on planning, controlling, and organizing, while leadership inspires, motivates, and drives change.

The distinction between management and leadership becomes increasingly relevant in a rapidly evolving workplace. Management is essential for maintaining order and executing established processes, but leadership is what drives organizations to embrace change and seize new opportunities. Leaders are tasked with navigating ambiguity and setting a course for progress, whereas managers ensure the efficient execution of this vision. The shift in workforce dynamics, particularly with the increasing inclusion of neurodivergent individuals, demands a nuanced understanding of leadership's role in talent development. Neurodivergent employees bring unique strengths and perspectives that can enhance innovation and problem-solving capabilities. However, these employees often require personalized approaches to development that align with their specific needs. Inclusive leadership approaches tailored to diverse learning styles and strengths are critical for organizations aiming to nurture high-performing teams.

Moreover, the inclusion of neurodivergent individuals challenges leaders and the function to rethink traditional models of personnel development. For example, a linear approach to standardized training may fail to address the unique contributions of neurodivergent employees effectively. Instead, leaders must adopt flexible, adaptive strategies that combine experiential learning opportunities with targeted coaching and mentorship. Such strategies encourage an inclusive environment where

M. Helmold, M. Martensen, *Leadership in Neurodiversity and Neurodivergence*, Diversity and Inclusion Research, https://doi.org/10.1007/978-3-032-09343-1_11

diverse talents can flourish, ensuring that the organization benefits from the full spectrum of skills and perspectives available.

Ultimately, leadership in personnel development extends beyond merely facilitating professional growth. It is about creating an environment where every individual feels valued and supported. Effective leaders understand that the success of their teams depends on their ability to foster trust, empower employees, and inspire collective commitment to organizational goals. This dual focus on individual and team development underscores the transformative potential of leadership in today's complex and diverse work environment.

11.2 Defining Management and Leadership

According to Kotter, management ensures the effective functioning of processes, emphasizing control, efficiency, and predictability. It is primarily focused on maintaining the status quo by managing resources, processes, and systems in a structured and consistent manner. Managers excel in enforcing discipline, setting short-term objectives, and ensuring operational excellence within defined parameters (Kotter, 1990). Conversely, leadership is centered around establishing direction, aligning people, and inspiring action to achieve a shared vision. Leaders are visionaries who challenge existing paradigms, drive organizational change, and motivate teams to pursue ambitious goals (Northouse, 2019). They focus on innovation, adaptability, and fostering a culture of collaboration and continuous improvement. Leadership is inherently dynamic, often requiring a deep understanding of human behavior and emotional intelligence to connect with and inspire teams effectively (Goleman et al., 2002). This distinction highlights how management and leadership, though interconnected, cater to different organizational needs. Management provides the foundation for stability and efficiency, while leadership propels growth, transformation, and long-term success. Table 11.1 illustrates these differences succinctly.

Managers ensure consistent competency building through structured training programs, creating a foundation of skills and operational excellence within teams. Structured programs provide employees with clear guidelines and measurable outcomes, fostering a predictable path to skill acquisition. Leaders, on the other hand, transcend these foundations by inspiring employees to exceed their potential, cultivating an environment of creativity and ownership. Through vision-setting and motivational strategies, leaders encourage individuals to think beyond the immediate tasks and explore innovative solutions to complex problems. Effective personnel development thus requires the dynamic interplay of both roles—managers ensuring

Table 11.1 Management vs. leadership (Kotter, 1990)

Aspect	Management	Leadership
Focus	Processes and systems	People and vision
Objective	Efficiency and consistency	Innovation and growth
Approach	Planning and controlling	Inspiring and motivating

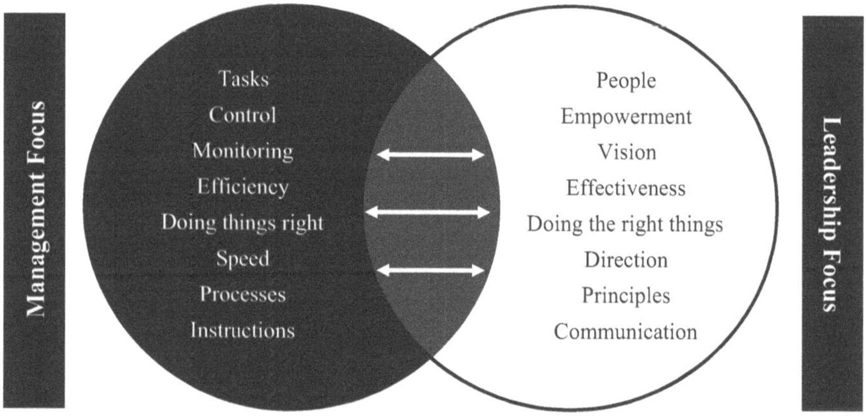

Fig. 11.1 Management and leadership focus

routine skill-building with precision and leaders driving innovation and engagement. This balance is particularly vital in diverse and neurodiverse workforces, where tailored approaches to learning and development can unlock the full spectrum of employee potential (Fig. 11.1).

11.3 Leadership Roles and Corresponding Developmental Strategies for Neurodiverse Teams

The origins of leadership roles in modern organizational frameworks can be traced to the interplay between classical management theories and evolving paradigms in behavioral science. Concepts such as task-oriented leadership, which aligns closely with the "Boss" role, emerged from studies like the Ohio State Leadership Studies (Stogdill, 1948). These studies highlighted the balance between initiating structure (task management) and consideration (team dynamics), laying the groundwork for categorizing leadership functions. Further refinement came with frameworks like Blake and Mouton's (1964) Managerial Grid, which introduced the dual focus on task and people-oriented leadership. Over time, scholars such as Trost (2020) expanded these foundational ideas to define specific roles that leaders undertake to manage and foster personnel development:

Boss: The boss directs tasks, monitors progress, and ensures compliance with organizational standards. This role is pivotal for maintaining accountability and operational discipline. For instance, a boss ensures team members meet their objectives efficiently while upholding quality benchmarks.

Enabler: The enabler removes barriers to success by providing resources and creating a supportive environment. Leaders in this role empower employees to harness their potential and overcome challenges. Enablers are instrumental in fostering a culture of collaboration and innovation.

Coach: A coach focuses on individual development, offering personalized feedback, motivation, and guidance. Coaches build on each employee's strengths and provide support to enhance performance. This role is particularly effective in helping employees achieve their long-term career goals.

Partner: The partner collaborates with employees as equals, fostering trust and inclusivity. This role emphasizes shared responsibility and mutual respect, which enhance team cohesion and organizational culture (Fig. 11.2).

In neurodiverse teams, adapting leadership roles is essential for meeting individual needs and unlocking the distinctive talents that each team member brings. Leadership in such environments requires a nuanced approach that not only recognizes individual strengths but also tailors strategies to overcome potential barriers. By understanding the interplay of task management, resource facilitation, and collaboration, leaders can create an inclusive culture that benefits all team members. Effective leadership in neurodiverse teams involves creating structured, yet flexible, systems that cater to diverse working styles. For instance, leaders need to assess the unique cognitive and emotional needs of their team members. Some employees may excel under well-defined workflows, while others thrive with opportunities for creative input and autonomy. Understanding these preferences is key to fostering engagement and productivity. Leaders should also integrate continuous feedback loops into their management practices. Regular check-ins and open communication channels ensure that employees feel supported and valued. Additionally, leveraging assistive technologies, such as project management tools or communication platforms, can streamline collaboration and minimize misunderstandings within diverse teams. Below are specific strategies tailored to maximize the potential of neurodivergent employees.

Boss: Leaders maintain accountability by establishing realistic and individualized goals. These goals must be clear and achievable to ensure that neurodivergent employees, who often thrive on predictability and structure, experience reduced stress and heightened clarity in their tasks.

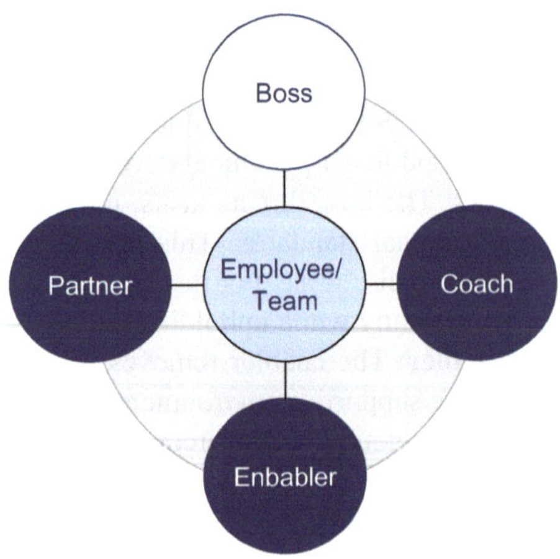

Fig. 11.2 Four leadership roles (Trost, 2020)

Table 11.2 Leadership roles and corresponding strategies for neurodiverse teams

Leadership role	Primary responsibility	Neurodiverse adaptation
Boss	Task management and oversight	Tailored goals, clear instructions, and predictable workflows
Enabler	Removing barriers	Sensory-friendly workspaces, customized tools, and flexible arrangements
Coach	Personalized development	Frequent feedback, structured coaching, and adaptive support for focused growth
Partner	Collaborative engagement	Inclusive decision-making and valuing diverse viewpoints to enhance team collaboration

Enabler: Creating supportive environments is critical for success. For example, providing sensory-friendly workspaces—such as those equipped with noise-canceling headphones, ergonomic furniture, or flexible seating options—empowers employees with autism spectrum disorders to perform effectively without unnecessary distractions or environmental discomfort.

Coach: Neurodivergent employees, such as those with ADHD, benefit greatly from structured development plans. Regular one-on-one check-ins and clearly defined milestones ensure that they remain focused and make measurable progress. A coach also provides motivational support to help employees overcome challenges and celebrate achievements.

Partner: Inclusive decision-making strengthens team cohesion and ensures that all voices are valued. Acting as a partner, a leader actively engages neurodivergent employees in collaborative problem-solving, leveraging their unique perspectives to drive creativity and innovation within the team (Table 11.2).

11.4 The 70:20:10 Rule of Learning and Development

11.4.1 Historical Background

The origins of the 70:20:10 learning model can be traced back to research conducted by McCall, Lombardo, and Eichinger at the Center for Creative Leadership in the 1980s. Their studies on leadership development highlighted that effective learning often occurs outside formal training programs, primarily through experiential and social interactions (McCall et al., 1988). This research revealed that leadership skills are best developed through real-world challenges and collaborative exchanges, where individuals refine their abilities by directly engaging with their environments.

This concept was later popularized by Michaels, Handfield-Jones, and Axelrod in their book *The War for Talent* (2001), which emphasized the importance of integrating structured training with experiential and social learning to cultivate high-performing leaders. The authors argued that traditional training methods, while valuable, often fail to prepare leaders for the complexities of dynamic and competitive environments. Instead, they advocated for a blended approach that incorporates hands-on projects, mentorship programs, and targeted education to holistically develop talent. Their work further underscored how such an integrated approach could enhance both individual and organizational performance, particularly in industries that demand innovation and adaptability.

The application of the 70:20:10 model extends far beyond leadership training. Its adaptability has made it a cornerstone of modern talent development strategies in various industries, enabling organizations to create scalable frameworks that address diverse learning needs. By focusing on experiential learning, the model allows employees to gain practical insights that complement theoretical knowledge, while the social learning component encourages collaboration and knowledge sharing across teams. Structured training, though comprising a smaller proportion, provides the foundational tools necessary to support the other two elements, ensuring a well-rounded development experience. As a result, the model has become a vital tool for organizations seeking to build resilient and agile workforces in today's complex global economy. Its scalability also makes it particularly effective in addressing workforce challenges posed by diverse teams, including those that are neurodivergent, by offering adaptable pathways for individualized development.

11.4.2 Overview of the Model and Leadership's Role in Supporting Each Dimension

As discussed, the 70:20:10 rule suggests that learning occurs across three primary dimensions: 70% of learning stems from real-world experiences and job-related challenges, such as problem-solving, decision-making, and leadership assignments; 20% originates from social interactions, including mentorship, coaching, and feedback-driven collaborations; and 10% comes from structured education programs like workshops, online courses, and formal training modules. Leaders as well as the HR function play a pivotal role in supporting each dimension (Fig. 11.3).

70% On-the-Job Learning: Leaders may give challenging assignments that push employees out of their comfort zones. This includes assigning responsibilities that require innovative thinking, decision-making under pressure, or cross-functional collaboration. For example, leaders might delegate tasks that involve launching a new initiative, managing a high-profile client project, or improving a failing process, enabling employees to develop advanced problem-solving skills. Additionally, leaders who encourage reflective practices help employees transform real-world experiences into valuable learning opportunities. Reflective exercises such as journaling or group debriefings can be used to analyze successes and identify areas for improvement, making experiential learning more impactful. These reflective methods not only strengthen learning but also enable employees to build self-awareness and resilience when facing similar challenges in the future.

20% Social Learning: Leaders foster a culture of mentorship and open communication. Creating structured peer-coaching programs or facilitating collaborative team projects enables employees to exchange knowledge and grow from

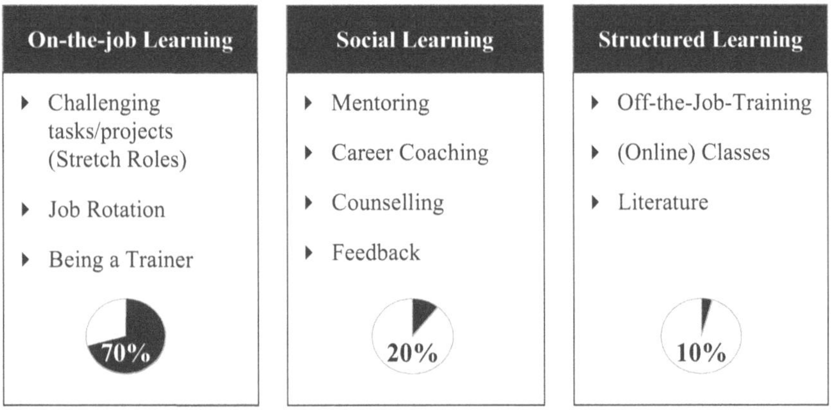

Fig. 11.3 The 70:20:10 rule of learning

interpersonal feedback. For instance, mentorship programs that pair experienced employees with newer team members can accelerate skill acquisition and knowledge transfer. Social learning can also occur during informal discussions, such as lunch-and-learns or brainstorming sessions, where ideas and insights are freely exchanged. Moreover, team-building activities and cross-departmental collaborations promote connectivity and strengthen trust among employees, enriching the social learning process. By embedding this collaborative culture into everyday operations, leaders can enhance team cohesion and innovation.

10% Structured Learning: While comprising a smaller share, formal training remains a vital component. Leaders who prioritize targeted workshops, certifications, or e-learning platforms align professional growth opportunities with individual career aspirations and organizational goals. Structured learning provides a theoretical foundation and introduces employees to best practices and emerging trends. Incorporating interactive formats, such as simulations or role-playing exercises, further enhances engagement and retention of knowledge. Additionally, leaders can leverage adaptive learning technologies that personalize training paths based on an employee's progress and competencies, ensuring that the structured learning experience is both relevant and effective. By combining theoretical rigor with innovative delivery methods, structured learning creates a robust framework that supports long-term career development.

11.4.3 Implications and Applications for Neurodivergent Employees

The 70:20:10 model serves as an effective framework for supporting neurodivergent employees due to its emphasis on adaptability and integration, making it easier to customize strategies for individual strengths and needs. Its structure allows leaders to create personalized approaches that not only address specific challenges but also amplify the unique abilities of neurodivergent individuals. By integrating experiential learning, social collaboration, and structured training, this model promotes a comprehensive development pathway. For instance, the adaptability of the framework makes it ideal for accommodating a wide range of cognitive and behavioral traits, ensuring that all employees feel supported in their growth. Additionally, the model encourages continuous feedback loops and iterative adjustments to development plans, enabling leaders to fine-tune their strategies as employees progress. This dynamic approach fosters an environment where neurodivergent employees can thrive, enhancing their contributions to organizational success while building skills tailored to their future aspirations.

70% On-the-Job Learning for Neurodivergent Employees: Leaders can craft experiential tasks tailored to leverage the distinct capabilities of neurodivergent employees. For instance, employees with autism spectrum disorders often excel in detail-oriented assignments, such as data analysis or quality assurance, while employees with ADHD thrive in fast-paced, dynamic roles that involve multitasking and problem-solving. To ensure effective engagement, leaders should provide structured workflows and clear expectations. Sensory-friendly work environments—incorporating noise-cancelling headphones, quiet zones, or adjustable lighting—can also reduce stress and help neurodivergent employees focus. Additionally, leaders should introduce opportunities for neurodivergent employees to engage in "stretch" projects that align with their unique strengths while also challenging them to expand their skill sets. For example, an employee excelling in attention to detail could be given a role in improving operational efficiency or conducting audits.

20% Social Learning for Neurodivergent Employees: Social learning opportunities can be enhanced by fostering inclusive mentorship and coaching relationships. Mentors who understand neurodivergent learning styles can guide employees through professional challenges, enabling them to thrive. Leaders can further support peer-to-peer learning by forming neurodiverse affinity groups or discussion forums. These initiatives encourage employees to share experiences, exchange strategies, and build a sense of community, leading to stronger interpersonal connections and mutual growth. Furthermore, team-based activities such as cross-departmental collaborations or group projects can facilitate interpersonal skills and introduce neurodivergent employees to diverse perspectives and workflows. Leaders should ensure that these activities are designed with clear goals and defined roles to prevent ambiguity and foster effective participation.

10% Structured Learning for Neurodivergent Employees: Formal training programs should integrate adaptive technologies and accessible resources to accommodate diverse needs. Video tutorials with closed captions, visual aids, and modular e-learning platforms can significantly enhance understanding for neurodivergent learners. Additionally, offering flexible schedules and pacing within structured programs ensures that employees can progress according to their individual comfort levels and learning speeds. This approach fosters a sense of accomplishment and long-term retention of knowledge. Structured learning can also include interactive simulations or gamified elements that make the content engaging and relevant.

Leaders can collaborate with instructional designers to ensure that training materials are inclusive and resonate with neurodivergent learners.

11.4.4 Advantages of the Model

The 70:20:10 model offers a versatile and effective approach to talent development that is particularly well-suited for modern workplaces. It provides a structured, yet flexible, framework that balances learning through experience, social interactions, and formal education. This balance ensures that employees receive well-rounded opportunities for growth while allowing organizations to adapt the model to meet the diverse needs of their workforce. For neurodivergent employees, in particular, this adaptability creates pathways for meaningful engagement and development, while fostering inclusivity across the organization. The following specific advantages can be identified.

Holistic Development: By emphasizing a balance of experiential, social, and formal learning, this model ensures comprehensive skill building, which is particularly beneficial for neurodivergent employees with varied learning preferences.

Employee Engagement: Neurodivergent employees feel more supported and engaged when learning opportunities cater to their unique strengths and challenges.

Inclusivity: Organizations can leverage the model to create inclusive talent development frameworks that recognize and celebrate neurodiverse capabilities.

Scalability: The model's adaptability makes it suitable for organizations of all sizes, allowing them to build diverse and dynamic teams capable of driving innovation and achieving strategic objectives.

11.5 Conclusion and Future Directions

Effective personnel development requires managers to establish structure and consistency while leaders inspire innovation and growth. By embracing Kotter's framework, Trost's leadership roles, and the 70:20:10 model, organizations can create a culture of continuous learning and development. Emphasizing inclusivity, particularly for neurodiverse workforces, strengthens this culture by ensuring that every employee has equitable access to development opportunities tailored to their unique strengths and challenges. Inclusivity in this context not only benefits employees but also drives organizational creativity and problem-solving by embracing diverse perspectives and skills.

To remain competitive and inclusive, organizations must focus on integrating advanced technologies and revising traditional frameworks. AI-driven learning platforms can deliver highly personalized training experiences, adapting to the varying speeds, preferences, and styles of neurodivergent employees. Such technologies make it possible to cater to individual needs while ensuring scalability and effectiveness. Additionally, fostering inclusivity requires a comprehensive redesign of talent development frameworks to address diverse cognitive needs. For example, integrating sensory-friendly environments, structured workflows, and accessible training materials ensures that all employees can participate fully in their development journey.

Leaders must go beyond basic compliance to actively create environments where neurodiverse talents can thrive. This requires a commitment to continuous learning and a deep understanding of the diverse needs within their teams. Adaptive leadership styles that focus on flexibility and empathy are crucial for addressing these needs effectively. By implementing innovative practices such as tailored mentorship programs, collaborative goal setting, and iterative feedback loops, leaders can align individual growth with broader organizational goals. Prioritizing inclusivity and adaptability allows organizations to unlock the full potential of their workforce, fostering resilience, engagement, and high performance across all levels. By cultivating such a culture, organizations position themselves as forward-thinking, inclusive, and prepared to navigate the complexities of today's global economy.

11.6 Case Study: Transforming Talent Development in a Multinational Firm

11.6.1 Context and Application of Concepts

A global technology firm faced challenges in retaining high-potential employees, fostering innovation, and supporting a neurodiverse workforce. By implementing leadership-driven development strategies, the organization transformed its approach to talent management. Leaders recognized that employee growth and engagement required a balance of structured programs, personalized support, and a culture of collaboration, with a particular focus on tailoring these approaches to the unique strengths and needs of neurodivergent employees. This balance was achieved by integrating experiential, social, and formal learning strategies into their talent management framework. By implementing leadership-driven development strategies, the organization transformed its approach to talent management. Leaders recognized that employee growth and engagement required a balance of structured programs, personalized support, and a culture of collaboration. This balance was achieved by integrating experiential, social, and formal learning strategies into their talent management framework. In particular, the following concepts were applied:

Management vs. Leadership: Operational managers ensured consistent training delivery by standardizing learning processes and providing employees with clear expectations. Meanwhile, leaders played a crucial role in inspiring innovation by encouraging creative problem-solving and supporting risk taking. Together, they created a dual-focused approach that balanced stability with growth.

Leadership Roles: Leaders acting as coaches provided tailored feedback and development plans to individuals, focusing on maximizing their unique strengths. The partnership role was also encouraged. Leaders worked alongside teams to co-create innovative solutions, fostering an environment of trust and collaboration that motivated employees to take ownership of their projects.

70:20:10 Model: Employees participated in stretch assignments (70%) that challenged them to take on new responsibilities, such as leading cross-functional projects. Specific accommodations were made for neurodivergent employees to ensure their success in these assignments. For example, tasks were broken down into smaller, manageable steps for individuals who benefit from high levels of structure, while sensory-friendly environments were provided to minimize distractions and support focus. Peer mentoring (20%) was established through dedicated mentoring networks, which paired junior employees with experienced staff for guidance and skill transfer. Neurodivergent employees were paired with mentors trained to understand their specific learning styles and communication preferences, creating a supportive and inclusive mentoring environment. Leadership workshops (10%) were customized to equip employees with critical skills for both individual and team success. These workshops included neurodiversity awareness training to help all employees and leaders understand how to leverage the unique strengths of their neurodivergent colleagues effectively.

11.6.2 Outcomes

Employee engagement scores improved by 30%, as measured through internal surveys. Employees reported feeling more connected to their work and empowered by the organization's emphasis on development. Also, attrition rates dropped by 25% over 2 years, attributed to employees feeling valued and supported in their career paths. Tremendous improvements were achieved in the area of innovation. Teams introduced 15 new product concepts, five of which successfully reached the market. Neurodivergent team members played a critical role in these successes, offering unique perspectives and innovative problem-solving approaches that significantly enhanced the company's product development processes. These innovations significantly enhanced the company's competitive position and demonstrated the effectiveness of integrating development-focused leadership practices.

References

Blake, R. R., & Mouton, J. S. (1964). *The managerial grid: The key to leadership excellence*. Gulf Publishing Company.

Goleman, D., Boyatzis, R., & McKee, A. (2002). *Primal leadership: Realizing the power of emotional intelligence*. Harvard Business Review Press.

Kotter, J. P. (1990). *A force for change: How leadership differs from management*. Free Press.

McCall, M. W., Lombardo, M. M., & Eichinger, R. W. (1988). *The lessons of experience: How successful executives develop on the job*. Lexington Books.

Michaels, E., Handfield-Jones, H., & Axelrod, B. (2001). *The war for talent*. Harvard Business School Press.

Northouse, P. G. (2019). *Leadership: Theory and practice*. Sage Publications.

Stogdill, R. M. (1948). Personal factors associated with leadership: A survey of the literature. *Journal of Psychology, 25*(1), 35–71.

Trost, A. (2020). *Human resources strategies: Balancing stability and agility in times of digitization*. Springer.

Neurodiversity and New Work 12

12.1 Embracing Neurodiversity Through New Work Principles

The concept of New Work has emerged as a transformative strategy for addressing the shortcomings of traditional work structures. At its core, it emphasizes autonomy, purpose, and flexibility as foundational elements of leadership and organizational practices. These principles aim to create work environments that not only prioritize humanity but also align deeply with individual values and aspirations. Notably, New Work serves as a comprehensive framework for integrating neurodiversity, empowering organizations to leverage the unique strengths of individuals with diverse cognitive profiles.

The term "New Work" was introduced by social philosopher Frithjof Bergmann in the late twentieth century. Bergmann's ideas stemmed from his critique of the industrial labor model, which, he argued, often led to alienation and diminished personal fulfillment. His vision centered on a work model that enabled individuals to engage in meaningful activities while balancing economic necessities with personal passions (Bergmann, 2004). This philosophy resonates strongly with neurodiversity initiatives, as inclusive and flexible workplaces allow neurodivergent individuals to thrive by maximizing their strengths and tailoring environments to their specific needs.

The relevance of New Work has become increasingly significant, particularly in the wake of the COVID-19 pandemic. The rapid adoption of remote work and flexible scheduling, facilitated by advancements in digital technologies, has disrupted traditional office-based employment and leadership models. These changes highlight the universal need for adaptability and flexibility, benefiting all employees while offering specific advantages to neurodivergent individuals seeking a balanced and productive work–life dynamic. Research supports this trend, showing that flexible work arrangements enhance both productivity and job satisfaction (Bloom

M. Helmold, M. Martensen, *Leadership in Neurodiversity and Neurodivergence*, Diversity and Inclusion Research, https://doi.org/10.1007/978-3-032-09343-1_12

et al., 2015). Furthermore, neurodiversity-focused programs within the New Work framework enable organizations to tap into underutilized talent pools, driving innovation and inclusivity.

New Work fosters an environment where neurodivergent employees can succeed through tailored leadership, task design, and communication strategies. For instance, flexible scheduling allows individuals with ADHD to work during their peak focus hours, while clear and structured task instructions reduce anxiety for employees on the autism spectrum. Purpose-driven roles further enable neurodivergent employees to align their unique abilities with organizational goals, fostering greater engagement and motivation. These roles not only enhance individual performance but also contribute to a dynamic and innovative organizational culture (Fig. 12.1).

Three key principles form the foundation of the New Work concept: autonomy, purpose, and flexibility. Autonomy grants employees the ability to decide how, when, and where they work, a factor that has been shown to significantly boost job satisfaction and engagement by allowing individuals greater control over their professional and personal lives (Ryan & Deci, 2000). For neurodivergent employees, autonomy is particularly critical, as it empowers them to tailor their workflows and environments to align better with their cognitive and sensory needs, enhancing both productivity and comfort. This principle also stimulates creativity by eliminating rigid structures that can inhibit optimal performance. Organizations that emphasize autonomy often report higher employee retention rates and more cohesive team dynamics, especially within diverse workforces.

Purpose, the second principle, involves aligning work with an individual's values and long-term goals. Such alignment not only fosters motivation but also cultivates a deeper sense of fulfillment. Companies which incorporate purpose into their operational models have observed elevated levels of employee engagement and commitment (Wrzesniewski & Dutton, 2001). For neurodivergent workers, purpose takes on added significance when roles are designed to leverage their distinct abilities—such as heightened analytical skills, creativity, or meticulous attention to detail.

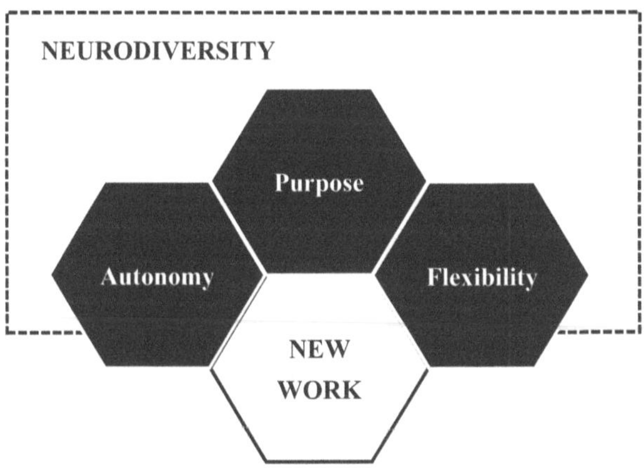

Fig. 12.1 Integrating New Work principles

Purpose-driven work frequently facilitates collaboration, as diverse perspectives merge to address complex challenges. This principle is instrumental in creating inclusive workplace cultures, ensuring that each employee finds meaning and value in their contributions.

The third principle, flexibility, refers to adaptable work arrangements, including flexible hours, remote work options, and fluid roles. Flexibility helps employees manage their responsibilities in ways that align with their personal and professional circumstances, promoting a healthier work–life balance. For neurodivergent individuals, flexibility often serves as a pathway to peak performance, as it reduces stress and enables more effective task management. Moreover, flexible arrangements are vital for improving workplace accessibility, making employment more inclusive for people with diverse needs. Digital tools like video conferencing and project management software have further highlighted the advantages of these arrangements (Bailey & Kurland, 2002). Hybrid work models, where employees split time between remote and in-office settings, have emerged as practical solutions for accommodating varied preferences and capabilities. Such models may enhance employee satisfaction, with neurodivergent workers benefiting substantially from these adaptive practices.

12.2 Historical and Technological Foundations of New Work

12.2.1 Insights from Bergmann and Drucker

In the 1980s, Frithjof Bergmann introduced a transformative critique of the industrial labor model, arguing that traditional work often resulted in alienation, lack of fulfillment, and a disconnect between individual aspirations and professional roles. His observations, particularly in Detroit during a period of economic hardship and technological changes, led him to propose a new vision for work (Bergmann, 1984). This vision centered around three key elements: autonomy, community, and meaningful engagement. Bergmann's philosophy emphasized that work should allow individuals to pursue creative and personally inspiring activities, rather than merely addressing economic needs. This concept, known as "New Work," seeks to restore purpose and dignity in the workplace. By promoting personalized workflows and fostering inclusivity, New Work allows organizations to better harness and celebrate diverse talents within their teams.

While not explicitly part of the New Work movement, Peter Drucker—a prominent management theorist—profoundly influenced contemporary perspectives on work. Drucker's mid-twentieth-century analysis highlighted a shift from manual labor to knowledge-based work, where intellectual capabilities became central to productivity (Drucker, 1999). He advocated for decentralization, continuous learning, and employee autonomy, ideas that closely align with the principles of New Work. Drucker's insights continue to shape discussions about workplace inclusivity and adaptability, particularly in how organizations integrate neurodivergent individuals into leadership and decision-making roles. His work underscores the growing recognition of cognitive diversity as a critical asset for innovation and resilience.

The evolution of New Work has also been driven by significant socioeconomic and technological developments. The rise of the knowledge economy, globalization, and rapid technological advancements in the late twentieth and early twenty-first centuries disrupted traditional workplace structures. Adaptability and flexibility became essential, and digital technologies such as the Internet, mobile devices, and cloud computing emerged as vital tools for enabling New Work. These innovations broke down barriers of time and location, empowering individuals to customize their work environments (Spreitzer et al., 2017). Project management platforms, collaboration tools, and other software have greatly improved accessibility, offering neurodivergent individuals the opportunity to tailor professional settings to their specific needs. For instance, tools like Slack and Zoom facilitate asynchronous communication, reducing sensory and social pressures, while platforms like Trello simplify task organization through visual design—providing a more intuitive workflow for many users.

Additionally, cultural shifts prioritizing well-being and work–life balance have reinforced the importance of New Work principles. Movements promoting mental health awareness and inclusive hiring practices naturally align with this philosophy, as both highlight the benefits of supportive and equitable workplace environments. Advancements in organizational psychology and neuroscience further validate the idea that inclusive practices enhance outcomes for both individuals and organizations. Ultimately, New Work represents more than a response to changing economic landscapes; it's a progressive approach that thrives on cognitive diversity, ensuring that all employees can contribute meaningfully and reach their full potential.

12.2.2 New Work and Its More Recent Embrace of Neurodiversity

Since its inception, the concept of New Work has undergone significant evolution, transitioning from a theoretical framework into a widely adopted strategy for modern organizations. Initially, it was championed as an approach to make work more humane and fulfilling, addressing the alienation and inefficiency often found in traditional workplace models. Today, New Work principles manifest in the growing prevalence of remote work, the gig economy, and flexible work structures, offering innovative solutions to the challenges of contemporary employment while fostering inclusivity and adaptability.

A notable example of New Work in practice is seen in companies like Google and Atlassian, which have integrated these principles into their organizational culture. Google's famous "20% time" policy (encouraging employees to devote a fifth of their working hours to passion projects) highlights the importance of autonomy and purpose in driving creativity and motivation (Pink, 2009). This initiative has proven particularly effective for neurodivergent employees, who tend to excel when allowed to focus on tasks that align with their strengths. Similarly, Atlassian emphasizes team collaboration and cross-functional innovation, enabling employees to contribute meaningfully while aligning their roles with personal skills and aspirations.

The COVID-19 pandemic played a pivotal role in accelerating the adoption of New Work practices. During this period, remote work became a necessity, showcasing the benefits of hybrid models and asynchronous communication systems. These adaptations not only demonstrated the feasibility of flexible work arrangements but also highlighted their capacity to better support neurodivergent employees. By facilitating customized workflows, New Work principles foster environments where employees feel empowered and valued (Bloom et al., 2015). Moreover, the pandemic underscored the critical importance of psychological safety. Organizations that coupled mental health support with New Work practices saw improvements in employee resilience, engagement, and overall job satisfaction.

As discussed, the three foundational principles of New Work—autonomy, purpose, and flexibility—serve as a blueprint for constructing workplaces that address the needs of diverse teams. Autonomy empowers individuals to design their schedules and workflows, purpose ensures alignment with meaningful objectives, and flexibility accommodates personal and professional circumstances. When neurodivergent employees are actively integrated into these frameworks, organizations unlock greater creativity, enhance problem-solving capabilities, and cultivate a culture of inclusion. This approach not only promotes individual well-being but also drives organizational innovation and competitiveness in today's rapidly changing global economy.

12.3 Core Principles of New Work and Their Relevance to Neurodiversity

12.3.1 Autonomy

Autonomy is a cornerstone of New Work, emphasizing the importance of giving employees control over how, when, and where they work (Deci & Ryan, 1985). For neurodivergent employees, this principle is particularly empowering. Autonomy allows them to adapt their work environments and routines to align with their individual sensory, cognitive, or organizational preferences. For instance, a neurodivergent individual might choose a quieter workspace, utilize noise-canceling headphones, or schedule tasks during periods of peak focus. This ability to self-direct not only enhances productivity but also reduces stress, enabling employees to perform at their best (Ryan & Deci, 2000).

Moreover, autonomy fosters confidence and a sense of ownership over one's work, which is particularly beneficial for employees who may have previously struggled in rigid or inflexible environments. It empowers individuals to identify and capitalize on their strengths, such as innovative thinking, attention to detail, or creative problem-solving. For example, a neurodivergent employee with exceptional analytical skills might thrive when given the freedom to design workflows or approaches tailored to their methods of thinking.

Organizations that promote autonomy, such as Atlassian with its "ShipIt Days"—where employees dedicate time to passion projects—foster creativity and innovation (Pink, 2009). These environments are ideal for neurodivergent workers, who may thrive when allowed to focus on projects that align with their unique strengths and interests. Additionally, autonomy within teams can encourage collaboration, as individuals bring diverse perspectives and methods to problem-solving, enriching team dynamics and outcomes.

12.3.2 Purpose

Purpose is the alignment of an individual's work with their values and long-term aspirations. This connection fosters deeper motivation and fulfillment, particularly for neurodivergent employees, who often excel in roles that emphasize their unique strengths. For example, individuals with autism may thrive in positions requiring exceptional attention to detail or strong pattern recognition, while those with ADHD often flourish in dynamic, fast-paced environments that reward creativity and adaptability. Aligning personal values with professional responsibilities helps neurodivergent employees feel more engaged and valued, significantly enhancing their contributions to the workplace.

Organizations that integrate purpose into their operations inspire employees to connect their personal values with the company's mission (Wrzesniewski & Dutton, 2001; Grant, 2012). For neurodivergent employees, purpose-driven work not only boosts engagement but also creates opportunities to demonstrate their talents in meaningful ways. For instance, purpose-aligned projects, such as those addressing environmental sustainability or social equity, allow neurodivergent individuals to focus on tasks that complement their abilities, including analytical research, innovative design, or creative problem-solving.

Additionally, workplaces centered on purpose often provide targeted opportunities for skill development, which can be particularly beneficial for neurodivergent employees. These environments support employees in refining their abilities in ways that align with their personal strengths while advancing organizational objectives. For instance, a neurodivergent individual with a talent for systems thinking might excel in roles focusing on process optimization or developing innovative solutions. By nurturing these strengths, purpose-oriented workplaces foster a deeper connection between employees and their organizations, promoting both personal and organizational growth.

Leaders play a pivotal role in fostering this sense of purpose by clearly demonstrating how individual contributions align with the broader company mission. By actively emphasizing this alignment, leaders create a culture of inclusion and collaboration, empowering neurodivergent employees to fully realize their potential while making meaningful contributions to organizational success.

12.3.3 Flexibility

Flexibility encompasses various adaptive work arrangements, such as flexible hours, remote work, and role adaptability. For neurodivergent employees, flexibility plays a critical role in creating supportive and effective work environments. By allowing individuals to tailor their workload to suit unique needs—whether that means working from home to avoid sensory distractions or adjusting deadlines to accommodate specific processing times—flexibility removes barriers and significantly enhances productivity. This approach is particularly beneficial for individuals who thrive with structured autonomy, such as breaking larger tasks into manageable parts or scheduling periodic breaks to reduce mental fatigue.

Beyond improving productivity, flexibility supports a healthier work–life balance. For neurodivergent employees, working in ways that align with their cognitive and sensory needs not only reduces stress but also fosters long-term engagement. For instance, flexible schedules might allow employees with ADHD to harness hyperfocus during peak hours while ensuring regular breaks for cognitive recharge. Similarly, employees on the autism spectrum can adjust their work environments to reduce sensory overload, improving concentration and task completion.

The integration of digital collaboration tools like Slack, Trello, and Asana strengthens flexible work environments by enabling seamless communication and organization (Spreitzer et al., 2017). Neurodivergent employees often benefit from tools that allow for visual task tracking, deadline management, and customizable workflows. These features reduce miscommunication and improve alignment across teams. Additionally, asynchronous communication through these platforms enables employees to process information at their own pace and respond thoughtfully, maximizing the value of their contributions.

Flexibility also fosters a culture of trust and accountability by accommodating diverse work styles and schedules. Neurodivergent employees often find this empowering, as it allows them to leverage their strengths. For example, creative employees might prefer working outside traditional hours when inspiration strikes, while those with strong analytical abilities might thrive with uninterrupted periods dedicated to problem-solving and precision tasks. This adaptability not only benefits individual employees but also enhances team performance by ensuring that everyone can work at their best. Moreover, flexibility drives innovation by encouraging diverse approaches to collaboration and problem-solving. Teams that prioritize adaptable work arrangements provide space for individuals with unique cognitive perspectives to flourish. In brainstorming sessions or on complex projects, neurodivergent employees' distinctive thinking styles can spark new ideas and lead to groundbreaking solutions, enriching organizational outcomes.

By embedding flexibility alongside autonomy and purpose in workplace practices, organizations can align with the principles of New Work while creating inclusive environments that empower neurodivergent employees. This approach not only improves team performance and drives innovation but also signals a genuine commitment to diversity and inclusion. Such practices attract and retain top talent while fostering long-term success for both employees and organizations.

12.4 Challenges of New Work and Neurodiversity

While the New Work paradigm offers numerous benefits, it also presents several challenges that affect all employees, particularly neurodivergent individuals. Addressing these challenges is essential to ensuring that the principles of New Work—autonomy, purpose, and flexibility—are inclusive and effective for a diverse workforce (Fig. 12.2).

12.4.1 Managing Remote Teams

Managing remote or hybrid teams is a major challenge of New Work. For neurodivergent employees, virtual environments can reduce sensory distractions but may also present obstacles like unclear communication or feelings of isolation. For example, employees on the autism spectrum may find ambiguous instructions or unstructured virtual meetings overwhelming, highlighting the importance of precise communication and accessible digital tools (Bailey & Kurland, 2002). Leaders should prioritize setting clear agendas, providing detailed instructions, and offering asynchronous communication options to allow employees to contribute without the stress of real-time interactions. These strategies can create an inclusive environment where neurodivergent individuals feel engaged and supported.

Fig. 12.2 Challenges in integrating New Work principles and neurodivergent employees

12.4.2 Equitable Access to Resources

Another critical challenge is ensuring equitable access to resources. Many neurodivergent employees may lack quiet, sensory-friendly workspaces or specialized technology, such as assistive software. Financial constraints can further limit their ability to set up productive home offices. Organizations should consider providing subsidies for ergonomic tools, offering access to sensory-friendly office spaces, and tailoring solutions like noise-cancelling headphones or accessible workflows to individual needs. Collaborating with employees to identify these needs ensures that support is both practical and impactful.

12.4.3 Maintaining Culture and Inclusion

Maintaining a cohesive organizational culture in distributed teams is especially complex for neurodivergent employees, who may rely on explicit communication to feel integrated. A lack of face-to-face interaction can lead to feelings of exclusion. Leaders should promote a psychologically safe environment by fostering practices that value diverse communication styles and preferences. Virtual events with multiple engagement formats, breakout sessions, and regular neurodiversity training can help reduce unconscious biases and ensure that neurodivergent employees feel understood and valued.

12.4.4 Balancing Flexibility and Structure

While flexibility is a hallmark of New Work, some neurodivergent employees thrive with added structure. For example, employees with ADHD may benefit from segmented tasks, clear deadlines, and periodic reminders to stay organized, while those on the autism spectrum often prefer consistent routines to reduce anxiety. Managers can strike a balance by implementing tools like project management software, customizable workflows, and structured daily check-ins. Encouraging open discussions about preferred working styles can further align individual needs with team objectives.

12.4.5 Adapting Leadership Approaches

Leaders must adjust their management approaches to support decentralized and autonomous work structures. Traditional performance evaluations may overlook the unique strengths of neurodivergent employees, such as creativity or analytical thinking. Instead, leaders should adopt empathetic, individualized feedback methods, such as one-on-one discussions or visual aids, to communicate effectively. Neurodiversity training for managers can help them recognize and amplify the contributions of neurodivergent employees, ensuring equitable support and fostering team success.

12.4.6 Data Security and Over-Reliance on Technology

The reliance on digital tools in New Work raises concerns about usability and reliability. Neurodivergent employees who depend on assistive technologies may face barriers when tools fail or require frequent updates. To mitigate these issues, organizations should provide robust IT support, conduct regular training on new technologies, and establish redundancy plans to prevent disruptions. Feedback mechanisms can also help identify and resolve usability concerns, ensuring that technology serves all employees effectively.

By addressing these challenges, organizations can uphold the principles of New Work while fostering inclusivity for neurodivergent employees. Proactively implementing equitable practices, offering mentorship programs, and investing in tailored resources will help organizations unlock the full potential of a diverse workforce. A workplace culture that values and empowers all employees ensures long-term success and alignment with New Work values.

12.5 Case Study: InnoTech Solutions – Building an Inclusive Workplace Through New Work

InnoTech Solutions, a global technology firm specializing in artificial intelligence solutions, implemented the principles of New Work to create an inclusive environment for neurodivergent employees. The company recognized that leveraging neurodiverse talent could not only improve its organizational culture but also drive innovation and problem-solving.

12.5.1 Background

InnoTech faced challenges with high turnover rates and lower engagement among employees, particularly in teams responsible for creative problem-solving and software testing. After conducting an internal survey, the leadership team identified that traditional work structures, such as rigid schedules and centralized decision-making, were limiting the potential of their neurodivergent employees. In response, they restructured their approach by adopting the New Work principles of autonomy, purpose, and flexibility. Additionally, InnoTech worked to identify systemic barriers that hindered the contributions of neurodivergent employees, such as inaccessible communication channels and outdated task assignment processes.

12.5.2 Implementation

Flexible Scheduling: InnoTech introduced fully flexible work hours and hybrid work arrangements. Employees were encouraged to structure their day in ways that suited their cognitive and sensory needs. Neurodivergent employees, for example, were allowed to work during their peak productivity hours and reduce exposure to sensory overload by working remotely. Special focus groups were created to ensure that flexible policies remained inclusive and effective for all employees.

Purpose-Driven Roles: The company redefined job roles to align with individual strengths. For instance, employees with autism spectrum were assigned roles requiring meticulous attention to detail, such as software quality assurance, while employees with ADHD were encouraged to contribute to dynamic, fast-paced projects involving brainstorming and innovation. Additionally, cross-departmental workshops were held to explore ways of customizing job responsibilities to fit each employee's skill set.

Assistive Tools and Resources: InnoTech provided access to assistive technologies, such as noise-cancelling headphones, speech-to-text software, and task management software with visual aids. Quiet zones were designated within office spaces, and virtual collaboration tools allowed asynchronous contributions to accommodate varied working styles. Employee feedback mechanisms ensured that new tools and resources met the evolving needs of the workforce.

Neurodiversity Training for Managers: Leadership underwent extensive training to better understand neurodiversity and adapt management practices. Managers learned to provide clear instructions, use structured feedback methods, and recognize the unique contributions of neurodivergent employees. In addition, mentorship programs were established to support neurodivergent employees as they navigated their career paths within the organization.

Inclusive Recruitment Processes: InnoTech revised its recruitment practices to ensure that they were neurodiversity-friendly. Traditional interviews were replaced with project-based assessments that allowed candidates to demonstrate their skills in practical settings, reducing bias and creating a more equitable hiring process.

12.5.3 Results

The changes led to measurable benefits for both employees and the organization:

Increased Productivity: Teams with neurodivergent members reported a 25% increase in efficiency. For example, the software testing team identified system vulnerabilities faster due to their enhanced focus and attention to detail. A specialized team focusing on algorithm optimization also reported a 20% improvement in project completion time.

Enhanced Innovation: The company's innovation team, which included employees with ADHD, developed three new product prototypes in half the projected time by leveraging their creative and fast-paced working style. Cross-functional collaborations involving neurodivergent employees generated innovative solutions to long-standing technical challenges.

Higher Employee Satisfaction: Surveys showed a 40% improvement in overall employee satisfaction, with neurodivergent employees highlighting the positive impact of flexible scheduling and purpose-driven work. Employee satisfaction surveys also indicated increased feelings of belonging and support among neurotypical employees who benefited from the inclusive culture.

Improved Retention: Employee turnover rates decreased by 30%, particularly in roles redesigned to align with neurodivergent strengths. Exit interviews revealed that flexible work arrangements and personalized development opportunities played a significant role in encouraging employees to stay.

Strengthened Employer Brand: InnoTech's commitment to inclusivity attracted top-tier talent, including both neurodivergent and neurotypical candidates, enhancing the company's reputation as a forward-thinking employer.

References

Bailey, D. E., & Kurland, N. B. (2002). A review of telework research: Findings, new directions, and lessons for the study of modern work. *Journal of Organizational Behavior, 23*(4), 383–400.

Bergmann, F. (1984). The new work philosophy. : University of Michigan Press.

Bergmann, F. (2004). *New work, new culture: Work we want and a culture that strengthens us*. University of Michigan Press.

Bloom, N., Liang, J., Roberts, J., & Ying, Z. J. (2015). Does working from home work? Evidence from a Chinese experiment. *The Quarterly Journal of Economics, 130*(1), 165–218.

Deci, E. L., & Ryan, R. M. (1985). *Intrinsic motivation and self-determination in human behavior*. Springer Science & Business Media.

Drucker, P. F. (1999). *Management challenges for the 21st century*. Harper Business.

Grant, A. M. (2012). Leading with meaning: Beneficiary contact, prosocial impact, and the performance effects of transformational leadership. *Academy of Management Journal, 55*(2), 458–476.

Pink, D. H. (2009). *Drive: The surprising truth about what motivates us*. Riverhead Books.

Ryan, R. M., & Deci, E. L. (2000). Self-determination theory and the facilitation of intrinsic motivation, social development, and well-being. *American Psychologist, 55*(1), 68–78.

Spreitzer, G., Cameron, L., & Garrett, L. (2017). Alternative work arrangements: Two images of the new world of work. *Annual Review of Organizational Psychology and Organizational Behavior, 4*, 473–499.

Wrzesniewski, A., & Dutton, J. E. (2001). Crafting a job: Revisioning employees as active crafters of their work. *Academy of Management Review, 26*(2), 179–201.

13 Performance Booster for Neurodivergent Employees and Teams: New Leadership

13.1 Exploiting the Potential in Neurodiversity to Achieve Performance Excellence

Workplace neurodiversity is an approach that explores neurological conditions not as disabilities, but as natural variations of what is considered the human standard. Up to 20% of the world's population is estimated to be neurodiverse according to various studies. This term includes conditions such as autism, ADHD, dyslexia, and Tourette's syndrome. But following decades of stigma around people with neurodiverse conditions, it appears that employers' attitudes to this talented minority are changing (World Economic Forum, 2023).

Successfully exploiting the potential from a neurodivergent workforce and team can foster productivity in organizations and lead to a performance booster. Leadership plays a crucial role in embracing the opportunities from variety and diversity in their teams. A growing number of prominent companies have reformed their leadership concepts and HR processes in order to access neurodiverse talent; among them are SAP, Hewlett Packard Enterprise (HPE), Microsoft, Willis Towers Watson, Ford, and EY. Many others, including Caterpillar, Dell Technologies, Deloitte, IBM, JPMorgan Chase, and UBS, have start-up or exploratory efforts under way (Austin & Pisano, 2017).

Leaders must be trained and made aware of the advantages organizations can gain from diversity in the backgrounds, disciplinary training, gender, culture, and other individual qualities of employees. Benefits from neurodiversity are similar but more direct. Because neurodiverse people are wired differently from "neurotypical" people, they may bring new perspectives to a company's efforts to create or recognize value (Dennison, 2024).

M. Helmold, M. Martensen, *Leadership in Neurodiversity and Neurodivergence*, Diversity and Inclusion Research, https://doi.org/10.1007/978-3-032-09343-1_13

13.2 Advantages of Neurodiversity

There are several advantages and benefits from having neurodivergent teams.

- Innovation and creativity
- Technical, design, and creative strengths
- New ways to solve problems
- High levels of concentration
- Keen accuracy and ability to detect errors
- Strong recall of information and detailed factual knowledge
- Reliability and persistence

Figure 13.1 depicts the advantages and opportunities from neurodivergent teams. Employees with the ADHD are energetic, dynamic, and spontaneous. As a result, employees can be more curious, creative, imaginative, innovative, and inventive. Workforce members with ADHD tend to be out-of-the-box thinkers, with an approach that can be highly prized in the workplace.

Autistic employees are honest, visually perceiving, and consider work issues from a different viewpoint and angle. Autistic traits are very useful for problem-solving tasks, as it means they consider a broader range of possibilities and are less likely to make assumptions. This means autistic people can bring unique perspectives and solve problems differently within a workplace.

Employees with mental health issues are perceptive, compassionate, and dedicated. Supporting these employees will lead to excellent results of the teams. Promoting these employees will provide a sense of accomplishment and purpose,

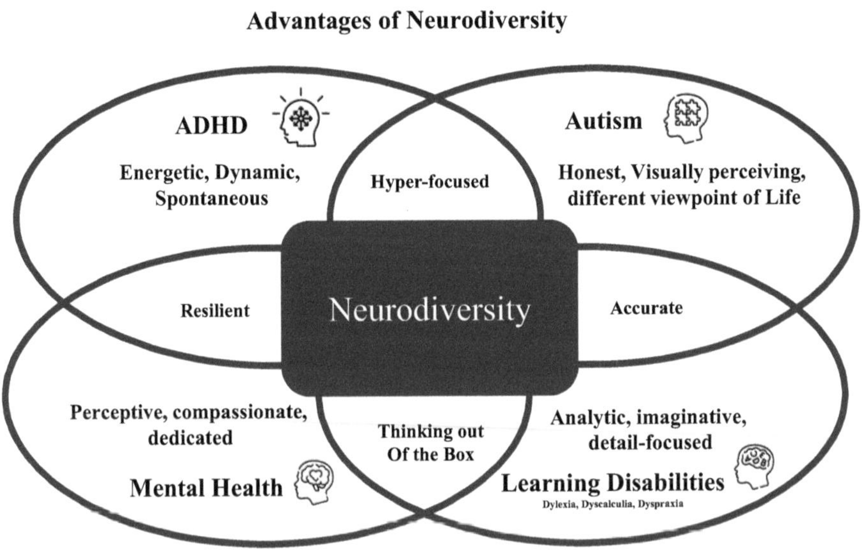

Fig. 13.1 Advantages from neurodivergent teams. Source: Author

boosting self-esteem and overall well-being so that the team will increase productivity.

Employees with certain learning disabilities, particularly autism and some cases of dyslexia, may show exceptional spatial reasoning and holistic visualization abilities. Usually, they are more analytic, imaginative, or detail-focused in comparison with other employees. Leaders can help employees to better understand their goals by breaking down the assignments into smaller tasks. Breaking down large assignments into smaller tasks with specific goals helps employees by making large projects more approachable, reducing the chance of procrastination, and providing a clear path to accomplishment. This method, also known as fragmentation, helps employees focus on immediate, manageable steps rather than being overwhelmed by the entire project. It also creates a clear sense of direction and allows for efficient resource allocation.

13.3 The New Leadership Triangle (E3 Model) for Performance Excellence of Neurodivergent Teams

Leadership has significantly changed and is no longer a static approach and behavior of mangers and corporate executives. Leadership is practiced every day in enterprises, businesses, in classrooms, in local clubs, on athletic fields, and yes, in homes. The new leadership paradigm embraces variety, diversity, and inclusion. Teams with and without employees of neurodivergent characteristics require a new style and way of leadership (Ternes and Wilke, 2018).

Required skills in the (new) leadership are the ability, the toolkit, and qualities needed to guide and influence others in a positive and successful way. Experts define effective leadership skills as finding ways to add value to others, focusing on growth and learning, doing the right thing, and staying disciplined.

Three major elements for leadership are empathy, empowerment, and engagement. Empathy can be described as the ability of leaders to take on the employee's perspective, to understand, to feel and possibly share and respond to his/her experience. Empathy includes social, cognitive, and emotional processes primarily concerned with understanding the workforce and employees. Empowerment is the degree of autonomy and self-determination in people and in communities. This enables them to represent their interests in a responsible and self-determined way, acting on their own authority. Engagement refers to the degree workers feel invested in, motivated by and passionate about their jobs and the company for which they work. The New Leadership Triangle model (E3 Model) integrates these three significant factors as shown in Fig. 13.2.

- Emotional empathy
- Education empowerment
- Ethical engagement

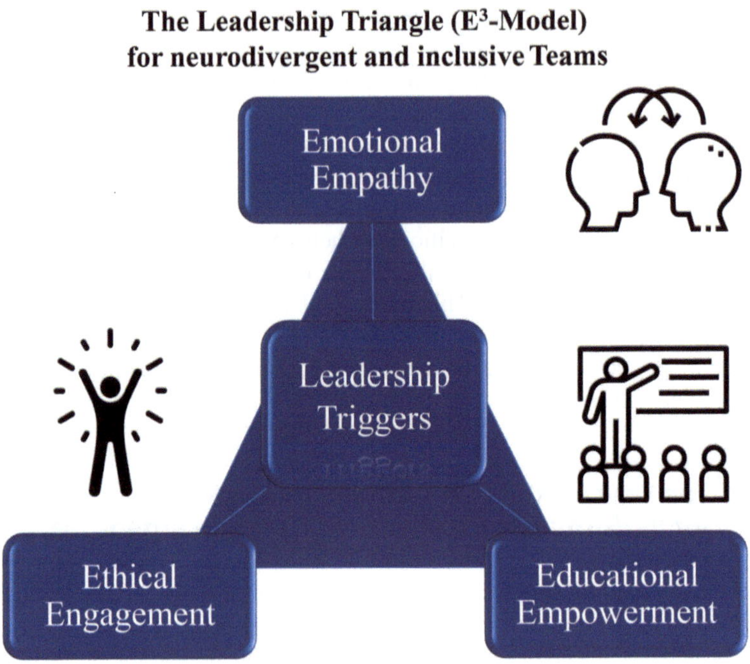

Fig. 13.2 The new leadership triangle for neurodivergent teams. Source: Author

13.4 Emotional Empathy

Leaders must understand how employees can be motivated and how to enable them to outperform; this refers to employees with and without neurodivergent characteristics. Therefore, it is necessary that they have a high level of emotional empathy. Emotional empathy is the practice of sharing an emotional experience with another person and truly understanding how they feel. This deep understanding of the other person's emotional state typically derives from similar experiences and is a key in the new leadership paradigm. When leaders show empathy in the workplace, they improve their effectiveness and increase trust and collaboration on their teams. Empathetic leadership is the key asset to organizations because it increases performance and culture. Emotional empathy is associated with emotional intelligence, but it goes beyond this competency (Helmold, 2022).

13.5 Educational Empowerment

Educational empowering is the process of new leaders of encouraging team members to manage themselves with a high degree of autonomy instead of supervising them directly. Empowerment leads to the fact that employees use their creativity and passion as shown in Fig. 13.2 (New Leadership Pyramid; Helmold et al., 2022).

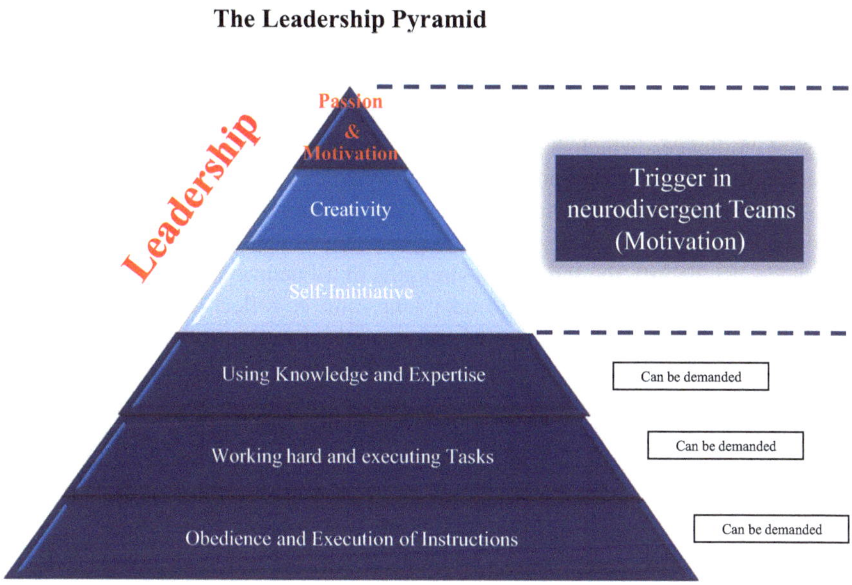

Fig. 13.3 Leadership pyramid: trigger creativity and passion in neurodivergent teams. Source: Author

Empowerment is a leadership element, which originated in the lean management approach (Helmold, 2022). In addition to this, it is important for new leaders to understand how to inspire team members in order to outperform. As a consequence, it is important, in which areas employees want and need to grow. The New Leadership Pyramid shows that leaders can demand obedience and execution of tasks. They can also force employees to work hard and use the acquired skills and expertise. But they cannot demand self-initiative, creativity, and passion. This is the reason why the lean management and new leadership approach is successful because it inspires team members to use their expertise, ability to think, and creativity. Educational empowerment shows significant increases in efficiency and productivity (Helmold et al., 2022) (Fig. 13.3).

13.6 Ethical Engagement

Ethical engagement refers to the leadership behavior that is directed by respect for ethical beliefs and values and for the dignity and rights of others. It is thus related to concepts such as trust, honesty, consideration, charisma, and fairness. Leaders should be open to neurodiverse employees and avoid an unconscious bias toward this group. New Leaders cultivate a positive and inspiring work environment by expressing clear ethical expectations, promoting open communication, and fostering a climate of trust and support. New Leaders have to be role models in applying ethical standards throughout their teams (Helmold, 2022).

13.7 Move from Competency Zone Toward Outperformance Zone

Leaders and their teams must have a development plan for neurotypical and neurodivergent employees, which is based on a proper status analysis, in order to achieve outperformance. One of the most important strategies for neurodivergent individuals is to know their strengths and embrace them (Titzmann, 2023). Neurodivergent individuals often have unique skills and perspectives that can bring value to the workplace. To identify the individual strengths of all neurotypical and neurodivergent staff members, leaders have to take some time to reflect on experiences and abilities (Helmold et al., 2023). In this context, it is crucial to reflect on competencies, learning, development, and growth in order to achieve outperformance as shown in Fig. 13.4.

The growing maturity of leaders, the awareness of diversity, and the development of specific skills in a neurotypical and neurodivergent environment for leaders are the key success factors to achieve high performance standards and excellence (Helmold, 2022; Frehner, 2023).

Whereas leader development is mainly about human capital development, leadership development focuses on social capital as the relevant outcome. Instead of focusing on the intrapersonal development of an individual leader as enhancing the potential for leadership, a collective focus recognizes that leadership is found in the interpersonal connections among individuals engaged in shared work. Rather than a systems input, leadership is conceptualized as an output shaped by systems structure and processes (Day, 2024).

The competency zone includes skills, knowledge, and expertise in which leaders and employees have a proper confidence and routine (Helmold et al., 2023). Leaders have to trigger the move beyond this zone into the learning zone.

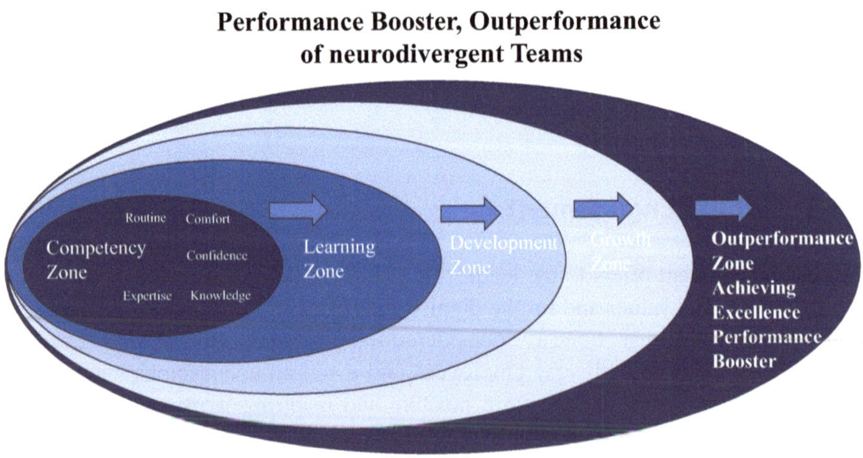

Fig. 13.4 Move toward outperformance in the new leadership paradigm. Source: Author

The learning zone is followed by the development zone, in which new skills, competencies and skills are acquired. Neurodiversity leaders must understand individual skill development by focusing on strengths and providing support for challenges, rather than applying a one-size-fits-all approach. This involves clear communication, flexible environments, and personalized plans that consider each employee's unique needs and how they can best contribute to the team.

This may lead to growth and new roles, behaviors, and routines (growth zone). As a result, teams will work more effectively, improve processes, and increase productivity by up to 50% (Helmold, 2022). Leaders must aim at inspiring their team members. It is necessary to be passionate, to be influential, and to rethink how to grow teams and individual employees. Leaders of neurodivergent teams must address and communicate effectively by challenging traditional instruments and finding new ways to communicate by adopting strategies like using clear, direct language, providing flexibility in communication methods, and fostering an inclusive environment through training and empathy. This involves shifting from expecting employees to conform to the organization to the organization adapting to the diverse needs of its employees. They must cultivate a positive communication culture. Leadership communication is a critical success factor of senior management teams and (new) leaders in the digital and human age to better interact and connect with others; drive innovation and adoption processes; and empower young minds with joy, abundance, and wisdom. In the classical view, leadership communication is part of management communication, which means leaders primarily use instruments focusing on teams, presentations, and negotiations. In the modern concept, new leaders are pioneers and game changers in communication (Pfeffermann & Schaller, 2023).

13.8 Case Study: Neurodiversity in HPE

13.8.1 Inclusive Culture

HPE is promoting neurodiversity. At HPE, the purpose is to advance the way people live and work. This is underpinned by a culture that believes in unconditional inclusion and the power of diversity to fuel innovation. The HPE Global Disabilities Network is the inclusive employee resource group with an explicit purpose of empowering team members who have a disability or have a connection to an individual with a disability. In partnership with their Accessibility Office, they host regular programming throughout the year on a range of topics addressing the needs of persons with disabilities and the organization to be disability-inclusive.

Some neurodiverse people are gifted with exceptional talents in mathematics, data analytics, or pattern recognition. HPE found that people with autism were up much faster at problem-solving, coding, or decoding. As a result, HPE has expanded cyber security teams and others to neurodivergent employees, as a neurodiverse candidate might be the best choice for the role. HPE encourages neurodiversity acceptance at work and find job candidates that will help the business grow (HPE, 2019). HPE took some special actions and measures for the recruitment of neurodivergent people.

13.8.2 Adjusting Recruitment Style

HPE has modified the searches to include neurodiverse jobseekers. HPE uses Artificial Intelligence (AI) to scan job applications. The AI technology helps not to filter out candidates who have resumes that are out of the norm.

13.8.3 Reconsidering the Interview Process in HPE

Instead of a face-to-face interview, HPE uses different question methods and types, which may make a neurodiverse job seeker not freeze up or come across as awkward, consider working through a typical project that allows them to showcase their skills and talents.

13.8.4 Onboarding Process for Neurodivergent Employees and a Pleasant Atmosphere

HPE has created special onboarding and welcome routines for neurodivergent newcomers (Austin & Pisano, 2017). Neurodiverse employees often need special accommodations such as private offices, which may frustrate other team members sitting in an open-plan space who'd like some privacy of their own. They may avoid office happy hours and appear antisocial. They might be blunt or off-putting in conversation. Leaders and Work Team members are trained and educated in the teammate's situation and feelings. Some integration programs assign a neurodiverse employee a "buddy" to occasionally check in on them or remind them to take breaks. Tell them they can come to you anytime a question or issue comes up.

Transparency and open lines of communication will keep everyone in the office comfortable, productive, and on track to succeed.

References

Austin, R. D., & Pisano, G. P. (2017). Neurodiversity as a competitive advantage—why you should embrace it in your workforce. *Harvard Business Review*. Retrieved February 1, 2025. https://hbr.org/2017/05/neurodiversity-as-a-competitive-advantage

Day, D. (2024). *Developing leaders and leadership. Principles, practices, and processes*. Springer.

Dennison, K. (2024). The benefits of neurodiversity in leadership. *Forbes*. Retrieved December 24, 2024. https://www.forbes.com/sites/karadennison/2024/11/14/the-benefits-of-neurodiversity-in-leadership/.

Frehner, T. (2023). *Führung heute: Erfolgsfaktoren des New Leadership*. Springer Gabler.

Helmold, M. (2022). Leadership. Agile, virtuelle und globale Führungskonzepte in Zeiten von neuen Megatrends. Springer .

Helmold, M., et al. (2022). *Lean Management, Kaizen, Kata and Keiretsu. Best-Practice Examples and Industry Insights from Japanese Concepts*. Springer.

Helmold, M. et al. (2023). New Work, Neues Arbeiten virtuell und in Präsenz. Konzepte und Werkzeuge zu innovativer, agiler und moderner Führung. Springer .

HPE. (2019). Neurodiversity in the Office. Retrieved December 27, 2024. https://www.hp.com/us-en/shop/tech-takes/neurodiversity-in-the-office.

Pfeffermann, N., & Schaller, M. (2023). *New leadership communication. Inspire your horizon.* Springer.

Siemens. (2024). InsideHR. How Siemens develops great leaders. Retrieved July 30, 2024. https://www.insidehr.com.au/how-siemens-develops-great-leaders/.

Ternes, A., & Wilke, C. D. (2018). *Agenda HR—Digitalisierung, Arbeit 4.0, New Leadership: Was Personalverantwortliche und Management jetzt nicht verpassen sollten.* Springer.

Titzmann, O. (2023). *New Leadership: Menschen führen heißt Menschen stärken.* Springer.

World Economic Forum (2023). Forum institutional. Employers now see neurodiversity as a strength in the workplace. Here's why. Retrieved December 24, 2024. https://www.weforum.org/stories/2023/05/neurodiversity-employers-workers-jobs/.

Neurodiversity: Unleashing the Potential from Neurodivergent Leadership 14

14.1 Benefits of Neurodivergent Leadership

Today's dynamic and competitive business landscape poses distinctive challenges for neurodivergent leaders. Despite holding mid- to senior-level positions, many find themselves at a career standstill.

The future of neurodivergent leadership is bright. As the world becomes more diverse, there is a growing need for leaders who can understand and appreciate different perspectives. Neurodivergent people are uniquely positioned to fill this need. With their unique strengths and perspectives, neurodivergent leaders can help to create a more inclusive and equitable world.

There are many advantages to having neurodivergent leaders (Pferffermann & Schaller, 2023; Helmold et al., 2022). Neurodivergent people often have unique strengths and perspectives that can be valuable in the workplace. For example, they may be more creative, innovative, and out-of-the-box thinkers. They may also be more empathetic and understanding of others.

Here are some of the specific benefits of neurodivergent leadership:

- Creativity and innovation: Neurodivergent people often have a unique way of thinking that can lead to new and innovative ideas. They may be able to see problems from a different perspective and come up with creative solutions.
- Out-of-the-box thinking: Neurodivergent people are often not afraid to think outside the box. They may be able to come up with new and creative solutions to problems that others have not been able to solve.
- Empathy and understanding: Neurodivergent people often have a strong understanding of others. They may be able to see the world from different perspectives and understand the needs of others.

M. Helmold, M. Martensen, *Leadership in Neurodiversity and Neurodivergence*, Diversity and Inclusion Research, https://doi.org/10.1007/978-3-032-09343-1_14

- Diversity and inclusion: Neurodivergent people can help to create a more diverse and inclusive workplace. They may be able to understand the needs of different people and create a workplace that is welcoming to everyone.

14.2 Tips for Neurodivergent Leaders

There are effective strategies to help neurodivergent leaders to excel in their roles while supporting their teams. These strategies are pertinent for neurodivergent leaders themselves as well as those tasked with supporting leaders and managers.

Unique strengths often found in neurodivergent leaders include the following:

- Diversity of thought, innovation, and creativity—neurodivergent leaders bring fresh perspectives and ideas, fostering revolutionary leaps in thinking.
- The ability to identify patterns and think associatively, a hallmark of neurodivergent thinking, leads to insightful solutions and effective decision-making.
- Applying forensic logic, analysis, and meticulous methodology to assess risks and opportunities.
- Neurodivergent individuals are known for their unwavering loyalty, commitment, and perseverance, which, research indicates, enhance overall team performance.
- Understanding what motivates and engages people is a critical leadership skill. Neurodivergent leaders often exhibit high levels of empathy and compassion, fostering a positive team culture.

14.3 Understanding Neurodiversity in the Workplace

14.3.1 Finding the Right Means

Neurodiversity refers to the different ways the brain functions and interprets information. Neurodivergent leaders often exhibit sharp contrasts in abilities, excelling in some areas, while struggling in others compared to their neurotypical colleagues. Unfortunately, these inconsistencies can be misunderstood as carelessness or indifference, hindering credibility and emotional regulation (Colella & Bruyère, 2022).

The more organizations can recognize their neurodivergent staff as different, not deficient, the easier it will be to identify where those individuals can be supported to reach their potential.

For leaders, it is important to identify the right measures in their teams in achieving the ultimate results and performance excellence (Helmold, 2022). A good mix of teams including the embracing of neurodivergent employees will lead to a competitive advantage (Austin & Pisano, 2017). The role of a leader in managing a diverse team is to maximize opportunities and minimize obstacles in the workplace.

The most effective way to achieve this is for a leader to use a range of strategies. Among these strategies are collaboration, accountability, judgment, a neutral lens, and leading by example (Helmold et al., 2023). Modern and new leadership styles combine diversity and inclusion in today's world. Diversity refers to the differences in people's backgrounds, experiences, and characteristics such as gender, race, age, religion, and sexual orientation. On the other hand, inclusion refers to the active and intentional creation of a culture where everyone feels welcome, valued, and respected, regardless of their differences (Helmold et al., 2025). Inclusive leaders create an environment where people can bring their whole selves to work and feel comfortable expressing their opinions and ideas.

Inclusive leadership is essential in creating a diverse and inclusive workplace and environment. It requires leaders to be intentional about creating a culture where everyone feels welcome and valued, regardless of their differences. Inclusive leaders actively seek out diverse perspectives, opinions, and experiences, and encourage employees to share their ideas (Dennison, 2024). They create an environment where people feel comfortable expressing their opinions and views, leading to increased engagement and productivity.

Figure 14.1 depicts recommendations for leaders in creating an inclusive environment and workspace for neurodivergent teams.

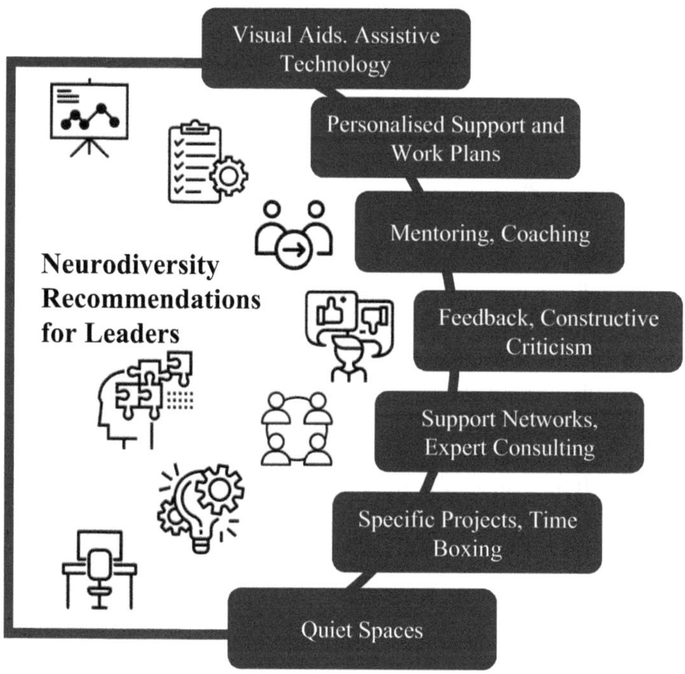

Fig. 14.1 Neurodiversity recommendations for leaders. Source: Author

14.3.2 Visual Aids and Assistive Technology

Visual supports, such as schedules, timers, and visual cues, can benefit neurodiverse students, particularly those with autism. These aids help students understand routines and instructions better, reducing anxiety and improving focus.

14.3.3 Personalized Support and Work Plans

Planning can be a challenging task for anyone, but it can be especially difficult for neurodivergent people. Neurodivergent people have different ways of processing information, thinking, and learning, which can make traditional planning strategies ineffective. However, with the right planning strategies, neurodivergent people can overcome these challenges and achieve their goals. One effective planning strategy for neurodivergent people is to break down large tasks into smaller, more manageable ones. This approach can help reduce overwhelm and make the task seem less daunting. Breaking tasks down into smaller steps can help neurodivergent people better understand what needs to be done and how to do it (Schroeder, 2024).

Neurodivergent workers may benefit from clear priorities and developed habits and routines. Well-structured work environments can include assisting with organizing priorities for work tasks, breaking tasks into smaller steps, and giving clear guidelines and times for breaks and lunches (Santuzzi & Patton, 2024). A systematic work plan is an execution strategy that helps in achieving different key results. It comprises specific steps to complete a project, initiative, and activities that could help in achieving the desired results.

14.3.4 Mentoring, Coaching

Neurodiversity is variable and can be complex, but is often invisible. Neurodivergent individuals think differently, but in doing so, they can make a valuable contribution to improving innovation and creativity within the workplace. They often possess outstanding, unique abilities in specific areas, but can find other aspects of the workplace and their role more difficult. Often, in mentoring or coaching, we focus on developing strengths and overcoming weaknesses (Dennison, 2024). When working with a neurodivergent individual, it is important to remember that their trait is not a weakness but a different way of doing things. Hence, your focus should shift from overcoming weaknesses to supporting the individual to use their skills, talents, and abilities to reach their full potential and realize their goals.

14.3.5 Feedback, Constructive Criticism

Providing feedback to employees is an essential part of any workplace, as it helps individuals grow and improve their performance. When it comes to neurodivergent employees, it's crucial for employers to approach this process with care and sensitivity (Schroeder, 2024). Neurodivergent individuals, such as those with autism, ADHD, or other neurological differences, may experience the world differently, and their communication styles and needs may vary.

One of the key aspects of giving feedback to neurodivergent employees is recognizing that their needs and preferences may vary widely. Some individuals may thrive on specific, constructive feedback, while others may prefer feedback in a more private and structured setting (Day, 2024).

An effective way for a manager to understand a neurodivergent employee's individual needs and preferences is to conduct a one-on-one meeting or discussion with the employee. During this conversation, the manager can ask open-ended questions and actively listen to the employee's responses (Frehner, 2023).

14.3.6 Support Network, Expert Consulting

Consultancies on neurodiversity provide neurodiversity support services to organizations around the globe. Neurodiversity consultancy services help inclusive organizations to understand:

- The benefits and challenges of having a neurodiverse and inclusive workforce, and how to overcome them
- How to approach workforce members with neurodivergent characteristics, and how to create a good and productive environment
- The importance of creating an inclusive culture and signposting employees to appropriate support, what support can be provided, and how this can maximize the potential of neurodivergent employees
- How they can make their recruitment process more inclusive and attractive to neurodiverse talent, and what typical strengths neurodivergent employees may have

14.3.7 Specific Projects, Time Boxing

Neurodivergent people often have difficulty with executive functioning skills, which can include time management, organization, and planning. Many neurodivergent individuals demonstrate an exceptional ability to focus intently on specific tasks. Specific project tasks should be accommodated for neurodivergent employees (Colella & Bruyère, 2022). In parallel, leaders must set clear priorities and a transparent structure. It is necessary to identify and prioritize tasks to focus on the most important activities first. Reduce Distractions: Create a quiet and organized environment in which to work in order to reduce distractions, improve focus and attention. Timeboxing allocates a maximum unit of time to an activity, called a timebox, within which a planned activity takes place. It is used by agile principles-based project management approaches and for personal time management (Helmold et al., 2023).

14.3.8 Quite Spaces and Sensory-Friendly Zones

Sensory sensitivities to noise are common among neurodiverse individuals. To create a focused room, incorporate sound-absorbing materials like acoustic panels, carpets, or curtains to reduce noise levels.

When designing a quiet space for neurodiverse employees, creating a safe and supportive environment that respects their sensory sensitivities and cognitive diversity is crucial (World Economic Forum, 2023). This can involve minimizing sensory distractions, providing sensory support tools (like noise-cancelling headphones or fidget tools), and using calming colors and lighting.

The goal of quiet spaces is to provide a retreat where individuals can relax, recharge, and escape the stimuli that may overwhelm them, helping neurodiverse individuals navigate the world more comfortably and safely. At work, quiet spaces are particularly valuable for students with sensory processing challenges. The following recommendations should be considered:

Sensory-Friendly Zones

- Low-Stimulation Environments: Designate areas with minimal sensory stimuli, such as soft lighting, muted colors, and soundproofing. These spaces are ideal for neurodiverse individuals who require a quiet environment for focused work.
- Quiet Rooms: Create dedicated quiet rooms where students can retreat for intense concentration or sensory breaks. These rooms should be equipped with comfortable seating and sensory-friendly lighting. Another option is installing acoustic or meeting pods to ensure a distraction and noise-free space.
- Sensory Break Areas: Design inviting sensory break areas with features like sensory toys, calming artwork, and adjustable lighting. These spaces offer neurodiverse students a place to decompress when sensory overload occurs.

Sensory-Friendly Lighting

- Lighting plays a crucial role in creating a comfortable environment for neurodiverse students. Harsh fluorescent lights or flickering fixtures can be overwhelming. Choose soft, adjustable lighting that minimizes glare and allows for individual control.
- Natural light sources are ideal; they promote a connection to the outdoors and a calming atmosphere.

Noise Control

- Sensory sensitivities to noise are common among neurodiverse individuals. To create a focused room, incorporate sound-absorbing materials like acoustic panels, carpets, or curtains to reduce noise levels.
- Provide noise-cancelling headphones for students who may need them to concentrate.

Quiet and Calm Spaces

- Designate specific areas within the room for quiet and calm activities. Comfortable seating, soft furnishings, and soothing colors can create a peaceful atmosphere.
- Consider creating small, private nooks where students can retreat for solitude or relaxation.

Flexible Furniture

- Neurodiverse students often have varying preferences for seating and workspace arrangements. Use modular and flexible furniture that can be easily adjusted to accommodate different needs.
- Ensure ample space for movement and different seating options, including standing desks and ergonomic chairs.

Sensory Tools and Resources

- Sensory tools like fidget toys, weighted blankets, or sensory-friendly textures can assist neurodiverse students in self-regulation. Providing these resources in a quiet room allows students to effectively manage sensory sensitivities and stress.

Color Palette

- Choose a calming and neutral color palette for the room to minimize visual distractions. Avoid overly bright or contrasting colors that may overwhelm some students. Soft, muted colors like blues, greens, or earth tones can create a serene atmosphere.

14.4 Case Study: Inclusive and Diversity Strategy in BMW

14.4.1 ConnectedWorks with Flexible and Quiet Workspaces

BMW AG employs a large number of people with disabilities. The company's quota of severely disabled people in 2023 was 6.0 percent, which—as in previous years—is above the legal requirements. In addition, BMW AG awarded contracts to workshops for people with disabilities last year for a total amount of around 42.7 million euros.

The BMW Group is developing new working environments for around 70,000 employees worldwide. With ConnectedWorks, the BMW Group is focusing on flexible working environments. Flexible, individual, sustainable—this is how the working environments at the BMW Group are designed. The company is gradually developing a future-oriented working environment for the employees worldwide in around 400 office buildings in over 40 countries (BMW, 2024). The BMW Group is thus continuing a process under the title "ConnectedWorks" that has been underway since 2017: For example, new forms of working models are being developed in the Munich Research and Innovation Center (FIZ) that offer every team a wide range of options for combining digital, location-independent formats with flexible, individually designed forms of collaboration on site. The aim is to achieve the best possible work results in each case. Creative communal areas are just as important as quiet zones in BMW. The design of these rooms with different heights and shapes of furniture makes employees comfortable.

14.4.2 Inclusion in BMW

There are visible limitations of many employees, who require a wheelchair or others aids. Many others are invisible like those of Peter Thiele, who has a cognitive impairment. In fact, there are a number of physical disabilities or mental disorders that are not immediately apparent from the outside: depression, Parkinson's, cancer and heart/lung diseases, but also panic attacks or bipolar disorders. Those affected do not always find it easy to talk about it. The BMW Group encourages them to do so. This is the only way they can receive protection, support, and encouragement. The representatives of severely disabled people at the respective locations are a contact point that encourages and supports people in this process.

Inclusion is implemented at the BMW Group worldwide, always taking into account the respective legal requirements and the different priorities of the individual nations. In Great Britain or Germany, for example, neurodiversity is a key issue, while in Australia, the focus is on mental health. Special action days are organized locally for this purpose (BMW, 2024). Despite all the different orientations, the core of the efforts remains the same: to enable all people to participate equally in working life.

14.4.3 Digitalization as Opportunity for Inclusion in BMW

Digitalization will open up new opportunities to drive inclusion even further. Technology barriers will be overcome and new activities can emerge in which people with disabilities can use their special strengths in a targeted manner. When it no longer makes a difference whether someone has a disability or not, inclusion has been achieved. Inclusion comes to life in the company when everyone participates: self-confident people with disabilities who deal with their limitations openly and naturally and show what skills and talents they have. And colleagues without disabilities who approach them without shyness or fear of contact. Inclusion requires encounters on an equal footing (BMW, 2024).

References

Austin, R. D., & Pisano, G. P. (2017). Neurodiversity as a competitive advantage—why you should embrace it in your workforce. *Harvard Business Review*. Retrieved February 1, 2025. https://hbr.org/2017/05/neurodiversity-as-a-competitive-advantage.

BMW. (2024). Handicap als Stärke. So gestaltet die BMW Group Inklusion. Retrieved December 27, 2024. https://www.press.bmwgroup.com/deutschland/article/detail/T0446355DE/handicap-als-staerke.

Colella, A., & Bruyère, S. M. (2022). *Taylor and Francis. Neurodiversity in the workplace. Interests, issues, and opportunities*. Taylor and Francis, Milton Park.

Day, D. (2024). *Developing leaders and leadership. Principles, practices, and processes*. Springer.

Dennison, K. (2024). The benefits of neurodiversity in leadership. *Forbes*. Retrieved December 24, 2024. https://www.forbes.com/sites/karadennison/2024/11/14/the-benefits-of-neurodiversity-in-leadership/.

Frehner, T. (2023). *Führung heute: Erfolgsfaktoren des New Leadership*. Springer Gabler.

Helmold, M. (2022). Leadership. Agile, virtuelle und globale Führungskonzepte in Zeiten von neuen Megatrends. Springer .

Helmold, M., et al. (2022). *Lean management, Kaizen, Kata and Keiretsu. Best-practice examples and industry insights from Japanese concepts*. Springer.

Helmold, M., et al. (2023). *New Work, Neues Arbeiten virtuell und in Präsenz. Konzepte und Werkzeuge zu innovativer, agiler und moderner Führung*. Springer.

Helmold, M., et al. (2025). New leadership. In *Innovative, lean and Agile leadership concepts*. World Science Publishing.

HPE (2019). Neurodiversity in the Office. Retrieved December 27, 2024. https://www.hp.com/us-en/shop/tech-takes/neurodiversity-in-the-office.

Pferffermann, N., & Schaller, M. (2023). *New leadership communication. Inspire your horizon*. Springer.

Santuzzi, A. M., & Patton, E. (2024). *Neurodiversity and work. Employment, identity, and support networks for neurominorities*. Springer Nature Switzerland. Imprint: Palgrave Macmillan Heidelberg.

Schroeder, S (2024). Effective planning strategies for neurodivergent people. Retrieved December 26, 2024. https://medium.com/@efbombcoach/effective-planning-strategies-for-neurodivergent-people-caff0d059091.

World Economic Forum. (2023). Forum institutional. Employers now see neurodiversity as a strength in the workplace. Here's why. Retrieved December 24, 2024. https://www.weforum.org/stories/2023/05/neurodiversity-employers-workers-jobs/.

MIX
Papier aus verantwortungsvollen Quellen
Paper from responsible sources
FSC® C105338

If you have any concerns about our products,
you can contact us on
ProductSafety@springernature.com

In case Publisher is established outside the EU,
the EU authorized representative is:
Springer Nature Customer Service Center GmbH
Europaplatz 3, 69115 Heidelberg, Germany

Printed by Libri Plureos GmbH
in Hamburg, Germany